T0320942

STOCHASTIC LINEAR
PROGRAMMING ALGORITHMS

Optimization Theory and Applications
A series of books and monographs on the theory and applications of optimization. Edited by K.-H. Elster † and F. Giannessi, University of Pisa, Italy

Volume 1

Stochastic Linear Programming Algorithms: A Comparison Based on a Model Management System
J. Mayer

This book is part of a series. The publisher will accept continuation orders which may be cancelled at any time and which provide for automatic billing and shipping of each title in the series upon publication. Please write for details.

STOCHASTIC LINEAR PROGRAMMING ALGORITHMS

A Comparison Based on a Model Management System

János Mayer

Gordon and Breach Science Publishers

Australia • Canada • China • France • Germany • India • Japan • Luxembourg
Malaysia • The Netherlands • Russia • Singapore • Switzerland • Thailand

Amsteldijk 166
1st Floor
1079 LH Amsterdam
The Netherlands

British Library Cataloguing in Publication Data

Mayer, János
    Stochastic linear programming algorithms: a comparison based on a model management.–
    (Optimization theory and applications; v. 1)
    1. Stochastic programming 2. Algorithms
    I. Title
    519.7'2

ISBN 90-5699-144-2
ISSN 1028-4168

To my family

CONTENTS

PREFACE

The subjects of stochastic programming are optimization models where some of the data are only known in the probabilistic sense, i.e. their stochastic dependency structure and probability distribution is specified. In stochastic linear programming (SLP) the underlying optimization model is a linear programming problem. There are various possibilities to formulate a meaningful optimization model in such a situation. In this book we consider two of these approaches: Two stage models and models with joint chance (or probabilistic) constraints. Our goal is to present some of the main solution approaches in a unified fashion, both from the theoretical and from the computational (empirical) point of view. The basis of the computational comparisons is SLP–IOR, a model management system for SLP, being under development by the author in cooperation with P. Kall.

The book is organized as follows: Chapter 1 provides the convex programming algorithmic background to the SLP algorithms presented in the book. The next Chapter introduces the two stage and chance constrained model classes. Chapter 3 is devoted to SLP solution algorithms followed by a Chapter summarizing the main points concerning implementation and describing the solvers. In Chapter 5 the testing environment is discussed which was used to obtain the computational results, presented in the last Chapter.

I wish to express my thanks first of all to Peter Kall for his continuous encouragement and many fruitful discussions. The excellent facilities and constructive seminars at his Institute provided very substantial contributions to this work. I am indebted to Jørgen Tind (University of Copenhagen) for drawing my attention to Wolsey decomposition, and to Kurt Marti (Federal Armed Forces University Munich) for his active interest in my work. Special thanks are due to my colleague Elisabeth Keller for her help in keeping my computing environment inperfect order. Finally, I thank my wife Ilona and my sons Tamás and András for their understanding and mental support.

This work was accepted as Habilitationsschrift by the Faculty of Economics of the University of Zurich in 1996 and has undergone only a minor formal revision.

Chapter 1

Algorithmic concepts in convex programming

In this chapter some general algorithms of convex programming will be presented. The presentation is restricted to those methods which either directly serve for solving an algebraic equivalent of an SLP[1] problem or provide an algorithmic framework for SLP algorithms. The first section summarizes the convex programming background of the algorithms. The next section is devoted to outer approximation methods followed by a section presenting the regularized counterparts of these methods. The subjects of Section 4 are barrier methods. Section 5 is devoted to central cutting plane methods. In the final section methods based on the reduced gradient idea are presented.

1.1 Optimality conditions and duality

In this section we briefly summarize some facts concerning optimality criteria and duality. Our purpose is to provide the convex programming basis for the algorithms which implies that only selected topics will be discussed. For a comprehensive treatment of the theory of convex programming see Rockafellar [136] .

We begin with duality. Let us consider the following mathematical programming problem:

$$\left.\begin{array}{c} \inf f\left(y\right) \\ G\left(y\right) \leq b \\ y \ \in Y \end{array}\right\} \tag{1.1}$$

where $Y \subseteq \mathbb{R}^{n}$; $f : Y \mapsto \mathbb{R}^{1}$; $G : Y \mapsto \mathbb{R}^{m}$; $G\left(y\right) = \left(\ G_{1}\left(y\right), \ldots, G_{m}\left(y\right)\ \right)^{\mathrm{T}}$. We assume that $Y \neq \emptyset$ is a compact convex set; f and G_{i}, $i = 1, \ldots, m$ are convex functions on Y, continuous relative to Y.

---

[1]SLP stands for Stochastic Linear Programming

1

For the algorithms discussed in subsequent sections we need strong duality. To assure this property a regularity condition is needed. We will assume that the following regularity condition holds:

Slater regularity: $\exists y°$, a feasible solution of (1.1), such that $y° \in \operatorname{ri} Y$ and $G_i(y°) < 0$ whenever $G_i(y)$ is nonlinear, $i = 1, \ldots, m$.

Let us remark that considering only strong duality, weaker regularity conditions would be sufficient, e.g. the Mangasarian-Fromovitz condition under the additional assumption of continuous differentiability of the functions involved in the model formulation (see e.g. McCormick [110]. Our motivation for choosing the Slater condition is twofold: Simplicity of presentation and the explicit need for a Slater point in most algorithms for jointly chance constrained problems. The compactness assumption concerning the set Y has been chosen for the sake of simplicity of presentation. It is definitely too strong for the linear case and will be dropped later on when we discuss linear models.

The Lagrange dual of problem (1.1) is the following (see e.g. [4]):

$$\sup_{u \geq 0} \inf_{y \in Y} [f(y) + u^T(G(y) - b)]. \tag{1.2}$$

By introducing the *dual function* $\gamma(u)$

$$\gamma(u) = \inf_{y \in Y} [f(y) + u^T(G(y) - b)] \tag{1.3}$$

it can be reformulated as

$$\sup_{u \geq 0} \gamma(u). \tag{1.4}$$

Notice that $\gamma(u)$ is trivially a concave function. The components of u will be called Lagrange multipliers.

Proposition 1.1 *(Duality theorem)*   *Under the conditions listed above we have:*

- *Weak duality:* Let y be a feasible solution of (1.1) and u be a feasible solution of (1.4). Then $\gamma(u) \leq f(y)$ holds.

- *Strong duality:* Let y^* be an optimal solution of (1.1). Then an optimal solution u^* of (1.4) exists and the optimal objective values are equal.

Proof: The weak duality is trivial, for the strong duality see e.g. Rockafellar [136], Corollary 28.3.1.   □

Next we will discuss conditions for the consistency of (1.1) by utilizing an idea of Geoffrion [47]. Clearly, (1.1) has a feasible solution iff the problem where the objective of (1.1) is replaced by the constant function $0^T y$, has an optimal solution. The Lagrange dual of this modified problem is the following:

$$\sup_{v \geq 0} \inf_{y \in Y} [v^T(G(y) - b)].$$

The dual function of the modified problem, denoted by $\delta(v)$,

$$\delta(v) = \inf_{y \in Y} \left[v^{\mathrm{T}} \left(G\left(y\right) - b \right) \right] \tag{1.5}$$

will be called in the sequel the *consistency function* of (1.1). Observe that in addition to its concavity $\delta(v)$ is also positively homogeneous. For the dual- and consistency functions the following obvious inequality holds:

$$\gamma(u) \geq \inf_{y \in Y} f(y) + \delta(u) \quad \forall u \geq 0. \tag{1.6}$$

The connection between the infeasibiliy of (1.1) and properties of the consistency function is formulated in the next proposition.

Proposition 1.2 *(Feasibility)* *Problem (1.1) is consistent $\Longleftrightarrow$* $\delta(v) \leq 0 \ \forall v \geq 0.$

Proof: The "$\Longrightarrow$" direction is trivial. The reverse follows immediately from the Bohnenblust-Karlin-Shapley theorem (see e.g. Rockafellar [136], Theorem 21.3), by utilizing the compactness of Y. $\qquad\square$

A vector v satisfying $v \geq 0$, $\delta(v) > 0$ will be called a *dual ray direction* for Problem (1.1). The previous proposition just states that (1.1) is inconsistent $\Longleftrightarrow$ there exists a dual ray direction. Under the compactness assumption the first term on the right hand side in (1.6) is finite so we observe: The existence of a dual ray direction implies that the objective $\gamma(u)$ of the Lagrange dual (1.4) is unbounded from above on the dual feasible domain.

The dual problem (1.2) can equivalently be reformulated as the following semi-infinite problem:

$$\left. \begin{array}{ll} \sup_{(u,u_0)} \left[-b^{\mathrm{T}} u + u_0 \right] & \\ f\left(y\right) + (G(y))^{\mathrm{T}} u - u_0 \ \geq 0, \quad \forall y \in Y & \\ \qquad\qquad u \qquad\quad \geq 0. & \end{array} \right\} \tag{1.7}$$

This view of the Lagrange dual opens the way to the Tind-Wolsey generalized duality for a broad class of mathematical programming problems including e.g. classes of discrete programming problems, see Tind and Wolsey [154]. Considering our purposes, this form of the Lagrange-dual is well-suited for pointing out some of the main features of generalizations of the Benders decomposition algorithm.

Notice that under our assumptions Problem (1.7) always has a feasible solution. Let us emphasize the following relations: u^* is an optimal solution of (1.2) $\Longleftrightarrow$ (1.7) has a finite optimum with an optimal solution (u^*, u_0^*), where

$$u_0^* = \inf_{y \in Y} \left[f\left(y\right) + (u^*)^{\mathrm{T}} G\left(y\right) \right].$$

Considering consistency, v is a dual ray direction for (1.1) $\Longleftrightarrow$ (v, v_0) fulfills the relations

$$\begin{array}{ll} (G\left(y\right))^{\mathrm{T}} v - v_0 \geq 0, \quad \forall y \in Y & \\ -b^{\mathrm{T}} v + v_0 \qquad > 0 & \end{array} \tag{1.8}$$

with $v_0 = \inf_{y \in Y} v^T G(y)$.

For later use let us also formulate the Kuhn-Tucker theorem. Our assumptions on Problem (1.1) will be modified as follows: Assume now that Y is an *open* convex set; f and the components of G are continuously differentiable; the other assumptions remain unchanged.

Proposition 1.3 *(Kuhn–Tucker theorem) Let x^* be a feasible solution of Problem (1.1) and $I(x^*) = \{\, i \mid G_i(x^*) = b_i \,\}$. x^* is an optimal solution of (1.1) iff*

$$\exists \lambda^* \geq 0 \text{ such that } \nabla f(x^*) + \sum_{i \in I(x^*)} \lambda_i^* \nabla G_i(x^*) = 0. \tag{1.9}$$

Proof: See e.g. [4].   □

The conditions above will be called the Kuhn-Tucker conditions; a pair of vectors (x^*, λ^*) fulfilling (1.9) will be called a Kuhn-Tucker point.

Notice that in (1.9) we may assume that the constraint gradients corresponding to positive λ_i^*'s are linearly independent. In fact, for fixed x^*, λ^* is a nonnegative solution of a system of linear equations. Such a solution can always be reduced to a basic feasible solution, see e.g. Kall [72].

Next we consider the following parametrized mathematical programming problem with $x \in X$ playing the role of a parameter:

$$\left. \begin{array}{l} Z(x) = \inf_y f(y) \\ \qquad\qquad G(y) \leq b - F(x) \\ \qquad\qquad y \in Y \end{array} \right\} \tag{1.10}$$

where $X \subseteq \mathbb{R}^{n_1}$; $Y \subseteq \mathbb{R}^{n_2}$; $f : Y \mapsto \mathbb{R}^1$; $G : Y \mapsto \mathbb{R}^m$; $F : X \mapsto \mathbb{R}^m$; $F(x) = (F_1(x), \ldots, F_m(x))^T$; $G(y) = (G_1(y), \ldots, G_m(y))^T$. We assume that $X \neq \emptyset$, $Y \neq \emptyset$; both sets are convex and compact; f, G_i, $i = 1, \ldots, m$ are convex functions, continuous relative to Y; F_i, $i = 1, \ldots, m$ are convex functions, continuous relative to X.

Let

$$X_1 = \{\, x \mid x \in X \text{ and } \exists y : G(y) \leq b - F(x), \, y \in Y \,\}. \tag{1.11}$$

We assume for the sake of simplicity of presentation that for any fixed $x \in X_1$, Problem (1.10) is Slater regular which is quite a strong assumption. In addition to the remarks we made in connection with Problem (1.1), let us mention a further reason for choosing this regularity condition. In some convergence proofs for outer approximation methods the assumption of locally uniformly bounded dual solutions is needed for which the Slater regularity is a necessary and sufficient condition (see Hogan [64]).

Notice that our assumptions imply that for each fixed $x \in X_1$, $Z(x)$ is finite, problem (1.10) has an optimal solution and the strong duality theorem (Proposition 1.1) holds.

Let us consider the dual function $\gamma : X \times \mathbb{R}^m_+ \mapsto \mathbb{R}^1$ corresponding to Problem (1.10):

$$\gamma(x, u) = \min_{y \in Y} [f(y) + u^T G(y)] - u^T(b - F(x)). \tag{1.12}$$

For any fixed u the first term in the sum above is independent of x. Let us denote it by $u_0(u)$, i.e.

$$u_0(u) = \min_{y \in Y} [f(y) + u^T G(y)] \tag{1.13}$$

yielding the representation

$$\gamma(x, u) = u_0(u) - u^T(b - F(x)). \tag{1.14}$$

Clearly, for any fixed $u \geq 0$, $\gamma(x, u)$ is a convex function on X. The support properties formulated in the next Proposition provide the basis for lower approximations to the optimal value function $Z(x)$ on the set X_1:

Proposition 1.4

1. *For any fixed $u^* \geq 0$ the following inequality holds:*

$$Z(x) \geq \gamma(x, u^*), \quad \forall x \in X_1.$$

2. *Let $x^* \in X_1$ and u^* be an optimal dual solution of (1.10), for $x = x^*$. Then we have:*

$$Z(x^*) = \gamma(x^*, u^*).$$

Proof: The assertions readily follow from the duality theorem. □

According to the previous proposition for any $x^* \in X_1$, and u^* being a corresponding optimal dual solution of (1.10) with $x = x^*$, the function $\gamma(x, u^*)$ supports $Z(x)$ from below at x^*. After some obvious reformulations and using the strong duality theorem we get the inequality below, which shows the supporting nature explicitly:

$$Z(x) \geq Z(x^*) + (u^*)^T(F(x) - F(x^*)), \quad \forall x \in X_1. \tag{1.15}$$

In the special case when $F(x) = Tx$, i.e. F is linear, we get

$$Z(x) \geq Z(x^*) + (u^*)^T T(x - x^*), \quad \forall x \in X_1 \tag{1.16}$$

which implies that $T^T u^* \in \partial Z(x^*)$ holds, i.e. $T^T u^*$ is a subgradient of $Z(x)$ at x^*.

Now we turn our attention to consistency problems. Let us consider the consistency function $\delta : X \times \mathbb{R}^m_+ \mapsto \mathbb{R}^1$. Similarly to the case of the dual function for any fixed $v \geq 0$ we get:

$$\delta(x, v) = v_0(v) - v^T(b - F(x)) \tag{1.17}$$

with

$$v_0(v) = \min_{y \in Y} v^T G(y) \tag{1.18}$$

showing that for any fixed $v \geq 0$, $\delta(x, v)$ is a convex function on X. The following separation property forms the basis for approximating the set X_1, the domain of finiteness for $Z(x)$, by sets determined through systems of inequalities.

Proposition 1.5

  1. *For any fixed $v^* \geq 0$ the following inequality holds:*

$$\delta(x, v^*) \leq 0, \quad \forall x \in X_1.$$

  2. *Let $x^* \in X \setminus X_1$ and v^* a dual ray direction for (1.10), with $x = x^*$. Then we have:*

$$\delta(x^*, v^*) > 0.$$

Proof: The assertions, including the existence of a dual ray direction, follow immediately from Proposition 1.2.   □

The semi-infinite Tind-Wolsey form of the dual of (1.10) is the following:

$$\left. \begin{array}{ll} \sup \left[-(\mathrm{b} - \mathrm{F}(x))^{\mathrm{T}} u + u_0 \right] & \\ \quad \mathrm{f}(y) + \mathrm{G}(y)^{\mathrm{T}} u - u_0 \geq 0, & \forall\, y \in Y \\ \qquad\qquad\qquad u \qquad \geq 0. & \end{array} \right\} \tag{1.19}$$

The feasible domain of (1.19) is independent of x which implies the following facts:

- (1.19) has a feasible solution for any $x \in X$.

- If (v, v_0) is a dual ray for $x = \hat{x}$ then it is a dual ray for any $x = \tilde{x}$ for which $-(\mathrm{b} - \mathrm{F}(\tilde{x}))^{\mathrm{T}} v + v_0 > 0$ holds, see relations (1.8).

1.2   Outer approximation methods

This section is devoted to outer approximation methods. The algorithms are presented as follows: We employ a general outer approximations scheme as a unifying framework. By gradually introducing structure into the problems under consideration, the various specific outer approximation algorithms are derived. With the exception of cutting plane methods, all algorithms presented in this section have their roots in the variable partitioning idea of Benders [5] .

In the first subsection the outer approximation scheme serving as our unifying framework will be presented. It is a special case of Wolsey's general resource decomposition algorithm [164]; we will call it the outer approximation scheme of Wolsey . The next subsection deals with cutting plane methods. The remaining subsections are devoted to outer approximation algorithms explicitly relying on the Benders [5] idea. In the third subsection convex programming problems having a separable structure in the Benders sense are considered. The generalized Benders decomposition of Geoffrion [48] , [49], as specialized to the separable structure, is the subject of this subsection. The case of a linear "second stage", i.e. the original Benders decomposition method, will be discussed next. The final subsection is devoted to discussing the special case of dual block-angular problems.

1.2.1 The outer approximation scheme of Wolsey

In this subsection a general mathematical programming problem will be formulated and the outer approximation scheme of Wolsey presented.

The following problem will be considered:

$$\left.\begin{array}{c} \min \ [f(x) + Z(x)] \\ x \in X \cap \tilde{X} \end{array}\right\} \tag{1.20}$$

where $X \subseteq \mathbb{R}^n$; $\tilde{X} \subseteq \mathbb{R}^n$; $f : X \mapsto \mathbb{R}^1$; $Z : \tilde{X} \mapsto \mathbb{R}^1$.

Let

$$X_1 = X \cap \tilde{X}$$

then our problem can be formulated as follows:

$$\left.\begin{array}{c} \min \ [f(x) + Z(x)] \\ x \in X_1. \end{array}\right\} \tag{1.21}$$

We will assume that $X_1 \neq \emptyset$; X is compact and X_1 is closed; the function f is continuous with respect to X and Z is continuous with respect to X_1.

These assumptions imply that problem (1.21) has a solution. Following Wolsey [164] the notions below will be introduced.

Definition 1.1 *(Support function) Let $x^* \in X_1$. The function $\sigma_{x^*} : X \mapsto \mathbb{R}^1$ is a support function to $Z(x)$ at x^* if*

1. $\sigma_{x^*}(x) \leq Z(x) \ \forall x \in X_1$,

2. $\sigma_{x^*}(x^*) = Z(x^*)$.

Definition 1.2 *(Cut function) Let $x^* \in X \setminus X_1$. The function $\tau_{x^*} : X \mapsto \mathbb{R}^1$ is a cut function of X_1 at x^* if*

1. $\tau_{x^*}(x) \leq 0 \ \forall x \in X_1$,

2. $\tau_{x^*}(x^*) > 0$.

We assume that for each $x^* \in X_1$ there exists a continuous support function $\sigma_{x^*}(x)$ and for each $x^* \in X \setminus X_1$ there exists a continuous cut function $\tau_{x^*}(x)$.

With the help of these functions (1.21) can equivalently be reformulated as the following semi-infinite problem:

$$\left.\begin{array}{ll} \min \; [\text{f}\,(x) \;\; + \max\{\sigma_{x^*}(x) \mid x^* \in X_1\}] \\ \qquad \tau_{x^*}(x) \quad\;\; \le 0 \;\; \forall x^* \in X \setminus X_1 \\ \qquad\quad x \quad\;\;\; \in X. \end{array}\right\} \tag{1.22}$$

The idea behind the problem formulation (1.21) and the assumption above is the following: The first term $\text{f}\,(x)$ in the objective of (1.21) is supposed to be an explicitly given "easy" function whereas the second term $Z\,(x)$ is considered as being "difficult", e.g. it is highly nonlinear or it is the optimal objective value function of an auxiliary optimization problem parametrized by x. Similarly for the constraints: X is assumed to be an "easy" set, e.g. a polyhedron, whereas $\tilde{X}$ is "difficult", e.g. it is determined by inequalities in terms of nonlinear functions or it is the domain where the auxiliary optimization problem determining $Z(x)$ has a finite optimum. The assumptions above imply that the original problem can be reformulated in such a way that the "difficult" parts are expressed in terms of support and cut functions. To make sense, the support and cut functions should be of a simpler nature as their "difficult" counterparts. The reformulated problem (1.22) is still difficult because it involves in general infinite families of functions but it opens the way for outer approximation strategies based on relaxation.

Problem (1.22) can equivalently be reformulated as follows:

$$\left.\begin{array}{ll} \min \quad \text{f}\,(x) + \eta \\ \quad \sigma_{x^*}(x) - \eta \le 0 \quad \forall x^* \in X_1 \\ \qquad \tau_{x^*}(x) \quad\;\;\, \le 0 \quad \forall x^* \in X \setminus X_1 \\ \qquad\quad x \quad\;\;\;\, \in X. \end{array}\right\} \tag{1.23}$$

The semi-infinite problem (1.23) is called the master problem. Let us summarize some obvious facts concerning this problem.

Clearly $X_1 = \{x \mid x \in X; \tau_{x^*}(x) \le 0, \; \forall x^* \in X \setminus X_1\}$ holds.

Let us consider an $\hat{x} \in X_1$. We have

$$\sigma_{x^*}(\hat{x}) \le Z\,(\hat{x}) = \sigma_{\hat{x}}\,(\hat{x}), \quad \forall x^* \in X_1.$$

As an immediate implication we have: For $\hat{x} \in X_1$ $(\hat{x}, \hat{\eta})$ is feasible for (1.23) iff it fulfills the constraint indexed by $\hat{x}$, i.e. if $\hat{\eta} \ge \sigma_{\hat{x}}\,(\hat{x}) = Z\,(\hat{x})$ holds. The minimal objective value for a fixed $\hat{x}$ and varying η is achieved when $\hat{\eta} = \sigma_{\hat{x}}\,(\hat{x})$ holds, i.e. the constraint indexed by $\hat{x}$ becomes active.

For the objective function of (1.23) let us introduce the following denotation:

$$\theta(x, \eta) = f\,(x) + \eta.$$

The common optimal objective value of (1.21), (1.22) and (1.23) will be denoted by θ^*.

Let

$$Z^p(x) = \max\{\sigma_{x_l}(x) \mid l = 1, \ldots, p\} \tag{1.24}$$

where $x_l \in X_1$, $l = 1, \ldots, p$.

The following relaxation of problem (1.22) will be considered:

$$\left.\begin{array}{ll} \min \mathrm{f}(x) + Z^p(x) & \\ \tau_{y_k}(x) & \leq 0, \quad k = 1, \ldots, q \\ x & \in X \end{array}\right\} \tag{1.25}$$

where $y_k \in X \setminus X_1$, $k = 1, \ldots, q$.

An evidently equivalent reformulation is the following:

$$\left.\begin{array}{ll} \min \quad \mathrm{f}(x) + \kappa \cdot \eta & \\ \sigma_{x_l}(x) - \eta \leq 0, & l = 1, \ldots, p \\ \tau_{y_k}(x) \leq 0, & k = 1, \ldots, q \\ x \in X \end{array}\right\} \tag{1.26}$$

where $x_l \in X_1$, $l = 1, \ldots, p$ and $y_k \in X \setminus X_1$, $k = 1, \ldots, q$ (the case when p=0 or q=0 means that support or cut functions are not present in the relaxed master problem, respectively), and $\kappa = 0$ for $p = 0$ whereas $\kappa = 1$ else.

Problem (1.26) is called the relaxed master problem.

Proposition 1.6 *Assume $p > 0$ and let $(\hat{x}, \hat{\eta})$ be an optimal solution of (1.26) with the property $\hat{x} \in X_1$ then the following inequalities hold:*

$$\mathrm{f}(\hat{x}) + \hat{\eta} = \theta(\hat{x}, \hat{\eta}) \leq \theta^* \leq \mathrm{f}(\hat{x}) + \sigma_{\hat{x}}(\hat{x}) \tag{1.27}$$

Proof: The inequalities follow readily from the relaxation property and from the feasibility of $\hat{x}$, respectively. □

We assume that an algorithm (also called an *oracle*) with the following capabilities is given.

Algorithm 1.0 *(Algorithm for computing support and cut functions)*

- For each $\hat{x} \in X$ it determines whether $\hat{x} \in X_1$ holds;

- if $\hat{x} \in X_1$ then it supplies a support function $\sigma_{\hat{x}}(x)$, otherwise a cut function $\tau_{\hat{x}}(x)$.

□

The general idea of the outer approximation scheme is the following:

Start with $p = 0$ and $q = 0$, i.e. with an empty set of support and cut constraints. Solve (1.26), the solution $\hat{x}$ will be the starting point of the algorithm. By utilizing Algorithm 1.0, successively add constraints to the relaxed master problem in an iteration loop. In each iteration solve the current relaxed master problem yielding the next $\hat{x}$. A constraint of the relaxed master involving a support function will be called an *optimality cut* whereas a constraint corresponding to a cut function will be called a *feasibility cut*.

The outer approximation method can now be formulated as follows.

Algorithm 1.1 *(Outer approximation scheme of Wolsey [164])*

 Step 0. *Initialize.*   Let p=0, q=0, determine $\hat{x}$ by solving (1.26), and set $\hat{\eta} = -\infty$. Choose a stopping tolerance $\epsilon^* > 0$.

 Step 1. *Solve a subproblem.*   Apply Algorithm 1.0 to $\hat{x}$. If $\hat{x} \in X_1$ then a support function $\sigma_{\hat{x}}(x)$ has been delivered, **Goto Step 3**; otherwise a cut function $\tau_{\hat{x}}(x)$ has been computed.

 Step 2. *Add a feasibility cut.*   Set $q := q + 1$; $y_q := \hat{x}$ and append $\tau_{y_q}(x) \leq 0$ to the constraint set of the relaxed master, **Goto Step 5**.

 Step 3. *Test for optimality.*   Check whether $\hat{\eta} \geq \sigma_{\hat{x}}(\hat{x}) - \epsilon^*$ holds. If yes then $\hat{x}$ is ϵ^*-optimal for (1.21) $\Longrightarrow$ **STOP**, otherwise:

 Step 4. *Add an optimality cut.*   Set $p := p+1$; $x_p := \hat{x}$ and append $\sigma_{x_p}(x) - \eta \leq 0$ to the constraint set of the relaxed master.

 Step 5. *Solve the new master problem.*   Solve the current relaxed master problem. Let the optimal solution be $(\hat{x}, \hat{\eta})$, with the following convention: As long as the relaxed master only contains feasibility cuts, return $\hat{\eta} = -\infty$. **Goto Step 1**.

$\square$

Steps 3 and 4 can also be interpreted as follows. We have solved the relaxed master problem. If the solution is feasible for the full master problem then it is obviously optimal. Otherwise we look for a mostly violated constraint of the full master at the point $(\hat{x}, \hat{\eta})$ and append this constraint to the relaxed master. In fact, the constraint indexed by $\hat{x}$ is evidently a mostly violated one at $(\hat{x}, \hat{\eta})$.

Let θ^l denote the optimal objective value of the relaxed master problem at iteration l. Then the following monotonicity relation obviously holds:

$$\theta^l \leq \theta^{l+1}, \; l = 1, 2, \ldots$$

Proposition 1.7 *The following two statements hold.*

  1. *None of the cut functions can reappear, i.e. $\tau_{\hat{x}}(x) \not\equiv \tau_{y_k}(x)$ on X, $k = 1, \ldots, q$.*

2. *If a support function reappears, i.e.* $\exists s\ (1 \leq s \leq p) :\ \sigma_{\hat{x}}(x) \equiv \sigma_{x_s}(x)$, *then the optimality criterion is fulfilled and consequently the algorithm stops.*

Proof: The reappearance of a cut function is excluded by the inequalities involving cut functions in (1.26). A reappearing support function clearly implies the fulfillment of the stopping criterion, see the corresponding inequality in (1.26).   □

Convergence behavior

At this generality nothing can be said about the convergence of the method. Nevertheless, as the following propositions show, the algorithm is well-defined, and if it stops then the optimal solution has been found.

Proposition 1.8 *The algorithm does not cycle.*

Proof: A feasibility cut clearly cuts off $\hat{x}$. An optimality cut cuts off $(\hat{x}, \hat{\eta})$; notice that $\hat{x}$ may reappear. In this case however the support function $\sigma_{\hat{x}}(x)$ also reappears which leads to stopping.   □

Proposition 1.9 *If the stopping criterion in Step 3 is fulfilled then $\hat{x}$ is an ϵ^*-optimal solution of (1.21).*

Proof: The assertion is an immediate consequence of the inequality (1.27).   □

Let us remark that in a recent paper Flippo and Rinnooy Kan [42] give a refinement of the Wolsey scheme. They also prove convergence under general assumptions concerning the optimal value function and point-to-set maps involved in the algorithm.

1.2.2   Cutting plane methods

This subsection is devoted to the cutting plane methods of Kelley [87] and Veinott [155] with two variants of the latter.

We consider nonlinear programming problems of the following type:

$$\left. \begin{array}{l} \min c^{\mathrm{T}} x \\[4pt] \quad g_i(x) \leq 0, \quad i = 1, \ldots, m \\[4pt] \quad x \in X \end{array} \right\} \qquad (1.28)$$

where $X \subseteq \mathbb{R}^n$; $c \in \mathbb{R}^n$; $g_i : X \mapsto \mathbb{R}^1, i = 1, \ldots, m$.

We assume that X is a nonempty bounded convex polyhedron; g_i, $i = 1, \ldots, m$ are convex and continuously differentiable and the feasible set is nonempty, i.e. an optimal solution exists.

In the problem formulation above we suppose that all of the functions g_i are nonlinear, i.e. linear constraints are incorporated into the defining inequalities of X. It is clear that problems with a convex nonlinear objective can be transformed into the form above. Let $\tilde{X} = \{ x \mid g_i(x) \leq 0,\ i = 1, \ldots, m \}$ and $Z(x) \equiv 0$. With this denotations our problem is a special case of the general problem (1.20) in the previous subsection. It is easy to see that the assumptions concerning (1.20) are fulfilled. In order to apply the Wolsey scheme the support and cut functions are to be specified. The support function for any $\hat{x}$ is trivially the constant zero function. The different cutting plane methods emerge through different choices of the cut functions. Observe that the algorithm will exclusively generate feasibility cuts. It will be stopped when ϵ-feasibility is achieved.

The key for getting cut functions is the convexity assumption. Let $\hat{x} \in X \setminus X_1$ and $i_0 :\ g_{i_0}(\hat{x}) = \max_i g_i(\hat{x})$ be the index of the mostly violated constraint. An obvious choice of a cut function is the following:

Construction of the cut function, Kelley cut

$$\tau_{\hat{x}}(x) = g_{i_0}(\hat{x}) + \nabla^T g_{i_0}(\hat{x})\,(x - \hat{x}). \tag{1.29}$$

Although in this case it is quite obvious, for the sake of uniform presentation we formulate the sub-algorithm for computing support and cut functions.

Algorithm 1.2 *(Algorithm for computing support and cut functions)*

Let $\epsilon > 0$ be a fixed feasibility tolerance. Check, whether $g_i(\hat{x}) \leq \epsilon,\ \forall i$ holds. If yes then return the constant zero support function, otherwise return the cut function defined by (1.29). □

Using this sub-algorithm we get the following cutting plane method:

Algorithm 1.3 *(Kelley's cutting plane method [87])*

The algorithm is defined by the outer approximation scheme 1.1 with the sub-algorithm 1.0 replaced by Algorithm 1.2. □

The cutting planes in Kelley's method are in general not supporting the feasible domain. Deeper cuts could be expected when employing supporting hyperplanes. For developing this idea we need an additional assumption:

Slater regularity: $\exists z^\circ \in X$ such that $g_i(z^\circ) < 0,\ \forall i$ holds.

Let $\hat{x} \in X \setminus X_1$ and z° be a Slater point. The line segment connecting z° and $\hat{x}$ intersects the boundary of the feasible domain at some point $\hat{y}$. The idea is to employ a cutting plane which supports the feasible domain at $\hat{y}$.

Algorithm 1.4 *(Algorithm for computing support and cut functions)*

Let $\epsilon > 0$ be a fixed feasibility tolerance. Check, whether $g_i(\hat{x}) \leq \epsilon$, $\forall i$ holds. If yes then return the constant zero support function, otherwise proceed as follows:

1. *(Linesearch)* Determine the intersection of the line segment joining z° and $\hat{x}$ with the boundary of the feasible domain. Let the intersection point be $\hat{y}$ and i_0 be an index for which $g_{i_0}(\hat{y}) = 0$ holds.

2. Return the cut function $\tau_{\hat{x}}(x) = \nabla^T g_{i_0}(\hat{y})(x - \hat{y})$. $\qquad\square$

This sub-algorithm leads to the following cutting plane method:

Algorithm 1.5 *(Veinott's supporting hyperplane method [155])*

The algorithm is defined by the outer approximation scheme 1.1 with the sub-algorithm 1.0 replaced by Algorithm 1.4. $\qquad\square$

In Veinott's method the Slater point is kept fixed. As it may be located quite unfavorably with respect to the optimal solution, this feature can slow down the convergence. In the following modifications the idea is to move the Slater point towards the current $\hat{y}$, along the line segment connecting z° and $\hat{x}$. This modified Slater point plays the role of z° in the supporting hyperplane method.

Let x^r be the solution of the master problem, y^r be the intersection point with the boundary and z^r be the Slater point, at iteration r, respectively. Let $z^1 = z^\circ$.

Zoutendijk's update [166]: Let $0 < \beta < 1$ be fixed and

$$z^{r+1} = z^r + \beta\left(y^r - z^r\right).$$

Szántai [151] reports on numerical difficulties originating in the constant rate of approaching the boundary. He suggests:

Szántai's update [151]:

$$z^{r+1} = z^r + \frac{1}{r+1}\left(y^r - z^r\right).$$

Observe that in the cutting plane methods summarized in this section the relaxed master problem is a linear programming problem.

Convergence behavior

Under the assumptions listed above all cutting plane methods considered here converge. This means that all accumulation points of the sequence of points, where the cutting planes are generated, are solutions of (1.28). This remains true when after solving the relaxed master problem inactive cuts are dropped. For a convergence proof see e.g. Luenberger [98].

1.2.3 Generalized Benders decomposition

The subject of this subsection is the generalized Benders decomposition algorithm of Geoffrion [48], [49],applied to problems having a separable structure as formulated below.

The following problem will be considered:

$$\left.\begin{array}{c} \min\left[f\left(x\right)+q\left(y\right)\right] \\ F\left(x\right)+G\left(y\right)\ \le b \\ x\qquad\ \in X \\ y\ \ \in Y \end{array}\right\} \tag{1.30}$$

where $f:\mathbb{R}^{n_1}\mapsto\mathbb{R}^1$; $q:\mathbb{R}^{n_2}\mapsto\mathbb{R}^1$; $F:\mathbb{R}^{n_1}\mapsto\mathbb{R}^m$; $G:\mathbb{R}^{n_2}\mapsto\mathbb{R}^m$; $F\left(x\right)=$
$\left(F_1\left(x\right),\ldots,F_m\left(x\right)\right)^{\mathrm{T}}$; $G\left(y\right)=\left(G_1\left(y\right),\ldots,G_m\left(y\right)\right)^{\mathrm{T}}$; $X\subseteq\mathbb{R}^{n_1}$, $Y\subseteq\mathbb{R}^{n_2}$.
We assume that f, q, F_i, G_i, $i=1,\ldots,m$, are convex functions; $X\neq\emptyset$, $Y\neq\emptyset$; both
sets are convex and compact and that (1.30) has a feasible solution. Let

$$X_1=\{\ x\ \mid\ x\in X\ \text{and}\ \exists y:\ G\left(y\right)\le b-F\left(x\right),\ y\in Y\ \} \tag{1.31}$$

and $Z:\ X_1\mapsto\mathbb{R}^1$ defined by

$$\left.\begin{array}{c} Z\left(x\right)=\min q\left(y\right) \\ G\left(y\right)\le b-F\left(x\right) \\ y\in Y. \end{array}\right\} \tag{1.32}$$

Our problem (1.30) can equivalently now be reformulated as

$$\left.\begin{array}{c} \min\left[f\left(x\right)+Z\left(x\right)\right] \\ x\in X_1 \end{array}\right\} \tag{1.33}$$

having precisely the form for which the outer approximation scheme of Wolsey has been
designed. For applying the Wolsey outer approximation scheme, we have to specify the
sub-algorithm which supplies support and cut functions.

We will assume Slater-regularity: Problem (1.32) is Slater regular, $\forall x\in X_1$. For the
motivation of this choice see Section 1.1.

The role of the hypothetical algorithm can be played by any convex programming method
which either supplies an optimal solution to the dual, or if the dual objective is unbounded
from above, a dual ray direction. Such an algorithm will be called *dual adequate*, see Geoffrion [49]. The support and cut functions can be obtained as follows.

Construction of the support function

Let $\hat{x} \in X_1$ and u^* be an optimal solution of the Lagrange dual of (1.32) with fixed $x = \hat{x}$, and consider

$$\sigma_{\hat{x}}(x) = \gamma(x, u^*) = -(\,b - F(x)\,)^T u^* + u_0^*(u^*) \tag{1.34}$$

with $u_0^*(u^*)$ defined by (1.13) for $u = u^*$ and taking into account (1.14). From Proposition 1.4 readily follows that the function $\sigma_{\hat{x}}(x)$ defined above is in fact a support function. Notice that $\hat{x}$ does not explicitly appear in the support function.

Construction of the cut function

Let $\hat{x} \in X \setminus X_1$ and v^* be a dual ray direction of (1.32) with $x = \hat{x}$ and consider

$$\tau_{\hat{x}}(x) = \delta(x, v^*) = -(\,b - F(x)\,)^T v^* + v_0^*(v^*) \tag{1.35}$$

where $v_0^*(v^*)$ is defined by (1.17) with $v = v^*$ and considering also (1.18). From Proposition 1.5 immediately follows that $\tau_{\hat{x}}(x)$ as defined above is a cut function. Notice that $\hat{x}$ does not explicitly appear in the cut function.

We may now specify the following algorithm:

Algorithm 1.6 *(Algorithm for computing support and cut functions)*

Apply a dual adequate convex programming algorithm to (1.32) with $x = \hat{x}$. If $x = \hat{x} \in X_1$ then compute a support function $\sigma_{\hat{x}}(x)$ according to (1.34), otherwise compute a cut function $\tau_{\hat{x}}(x)$ according to (1.35). $\quad\square$

The decomposition method is the following:

Algorithm 1.7 *(Decomposition method of Geoffrion [48],[49])*

The algorithm is defined by the outer approximation scheme 1.1 with the sub-algorithm 1.0 replaced by Algorithm 1.6. $\quad\square$

Convergence behavior

Proposition 1.10 *Assume that the number of subsequent feasibility cuts is finite throughout. Then the generalized Benders decomposition algorithm terminates in a finite number of iterations with an ϵ^*-optimal solution.*

Proof: See Geoffrion [49] and for an alternative proof Hogan [64], the latter also for cut-dropping variants of the method. Notice that the proof works with locally uniformly bounded dual solutions for which the Slater-regularity is a necessary and sufficient condition (see Hogan [64]). $\quad\square$

1.2.4   Benders decomposition

In this subsection the following problem will be considered:

$$\left.\begin{array}{rl} \min\,[\,f\,(x) + q^T y\,] & \\ F\,(x) + W\,y & \le b \\ x & \in X \\ y & \ge 0 \end{array}\right\} \tag{1.36}$$

where $f : \mathbb{R}^{n_1} \mapsto \mathbb{R}^1$; $F : \mathbb{R}^{n_1} \mapsto \mathbb{R}^m$; $q \in \mathbb{R}^{n_2}$; W is an $(m \times n_2)$ matrix and $F\,(x) = (\,F_1\,(x), \ldots, F_m\,(x)\,)^T$; $X \subseteq \mathbb{R}^{n_1}$.

All we need to assume is that (1.36) as well as all relaxed master problems (1.26) have solutions. This is fulfilled e.g. if (1.36) is feasible and X is a bounded finite set; the original Benders decomposition method has been designed for this mixed-integer problem type.

Considering a fixed $x \in X$ the convex programming problem (1.32) along with its dual is now an LP primal-dual pair:

$$\left.\begin{array}{rl} \min\,q^T y & \\ W\,y \le b - F\,(x) & \\ y \ge 0 & \end{array}\right\} \tag{1.37}$$

and

$$\left.\begin{array}{rl} \max\,[\,-(\,b - F\,(x)\,)^T u\,] & \\ W^T u & \ge -q, \\ u & \ge 0. \end{array}\right\} \tag{1.38}$$

Our assumption that (1.36) has a solution implies that there exists at least one $x \in X$ such that the primal LP (1.37) has a solution. From this immediately follows that the dual LP (1.38) is feasible for any $x \in X$ as its feasible domain does not depend on x. Notice that despite the fact that $Y = \{y \mid y \ge 0\}$ is a noncompact set, we have no problems in getting support and cut functions by LP-duality. The simplex method is obviously dual adequate.

Construction of the support function

Let u^* be an optimal basic solution of (1.38) with $x = \hat{x}$. As in the linear case $u_0^* = 0$ obviously holds, from (1.34) we get:

$$\sigma_{\hat{x}}(x) = -(\,b - F\,(x)\,)^T u^* = (u^*)^T F\,(x) - b^T u^*. \tag{1.39}$$

Construction of the cut function

Let v^* be an extremal direction of the polyhedral cone $\{v \mid W^T v \ge 0,\ v \ge 0\}$ with the property $-(\,b - F\,(\hat{x})\,)^T v^* > 0$. Such a v^* is delivered by the simplex method in the unbounded case. With $v_0^* = 0$ this is obviously a dual ray direction of (1.37) with $x = \hat{x}$. From (1.35) we get:

$$\tau_{\hat{x}}(x) = -(\,b - F(x)\,)^T v^* = (v^*)^T F(x) - b^T v^*. \tag{1.40}$$

Algorithm 1.8 *(Algorithm for computing support and cut functions)*

Apply the simplex method to (1.37) with $x = \hat{x}$. If $x = \hat{x} \in X_1$ then compute a support function $\sigma_{\hat{x}}(x)$ according to (1.39), otherwise compute a cut function $\tau_{\hat{x}}(x)$ according to (1.40). □

The classical Benders decomposition method is the following:

Algorithm 1.9 *(Decomposition method of Benders [5])*

The algorithm is defined by the outer approximation scheme 1.1 with the following modifications: The sub-algorithm 1.0 is replaced by Algorithm 1.8 and for the stopping tolerance $\epsilon^* = 0$ is also permitted thus allowing for an exact solution (constrained just by machine precision). □

Convergence behavior

Proposition 1.11 *The Benders decomposition method terminates in a finite number of steps with a solution of (1.36).*

Proof: Observe that the support and cut functions are completely determined by basic feasible solutions and extremal directions of the dual problem (1.38), respectively. Both of these sets being finite this implies that there exists only a finite number of different support and cut functions. In view of Proposition 1.7 it follows that the optimality criterion is fulfilled after a finite number of iterations. □

For the sake of completeness let us finally consider the completely linear case with the following structure:

$$\left. \begin{array}{rl} \min [\, c^T x + q^T y \,] & \\ A\,x & = b \\ T\,x + W\,y & = h \\ x & \geq 0 \\ y & \geq 0. \end{array} \right\} \tag{1.41}$$

We choose $X = \{\, x \mid A x = b, \; x \geq 0 \,\}$ and the support and cut functions specialize as:

Support function:

$$\sigma_{\hat{x}}(x) = (\,\mathrm{h} - \mathrm{T}x\,)^{\mathrm{T}}\, u^* = -(u^*)^{\mathrm{T}}\,\mathrm{T}\,x + (u^*)^{\mathrm{T}}\mathrm{h}. \qquad (1.42)$$

Cut function:

$$\tau_{\hat{x}}(x) = (\,\mathrm{h} - \mathrm{T}x\,)^{\mathrm{T}}\, v^* = -(v^*)^{\mathrm{T}}\,\mathrm{T}\,x + (v^*)^{\mathrm{T}}\mathrm{h}. \qquad (1.43)$$

1.2.5   Decomposition methods for dual block–angular problems

The basic ideas behind decomposition algorithms for dual block–angular problems can be best understood in the general terms of the Wolsey scheme. We consider the following problem:

$$\left.\begin{array}{c} \min\ [\,\mathrm{f}\,(x) + Z\,(x)\,] \\[4pt] x \in X \cap \tilde{X} \end{array}\right\} \qquad (1.44)$$

where now

$$Z(x) = \sum_{i=1}^{L} Z_i(x)$$

$$\tilde{X} = \bigcap_{i=1}^{L} \tilde{X}_i$$

'with $X \subseteq \mathrm{I\!R}^n$; $\tilde{X}_i \subseteq \mathrm{I\!R}^n$, $i = 1, \ldots, L$; $\mathrm{f} : X \mapsto \mathrm{I\!R}^1$; $Z_i : \tilde{X}_i \mapsto \mathrm{I\!R}^1$, $i = 1, \ldots, L$; $L > 0$. Let

$$X_i = X \cap \tilde{X}_i\ i = 1, \ldots, L; \ \text{ and } \ \hat{X} = \bigcap_{i=1}^{L} X_i$$

then our problem can be reformulated as:

$$\left.\begin{array}{c} \min\ [\,\mathrm{f}\,(x) + Z\,(x)\,] \\[4pt] x \in \hat{X}. \end{array}\right\} \qquad (1.45)$$

Our assumptions are as follows: $X_i \neq \emptyset$, $i = 1, \ldots, L$; X is compact and X_i, $i = 1, \ldots, L$ are closed. The function f is continuous with respect to X and Z_i is continuous with respect to X_i, $i = 1, \ldots, L$. For each $x^* \in X_i$ there exists a continuous support function $\sigma_{x^*}^i(x)$ to Z_i at x^*, $i = 1, \ldots, L$. For each $x^* \in X \setminus X_i$ there exists a continuous cut function $\tau_{x^*}^i(x)$ of X_i at x^*, $i = 1, \ldots, L$.

In complete analogy with the Wolsey scheme we hypothesize the availability of an algorithm (oracle):

Algorithm 1.10 *(Algorithm for computing support and cut functions)*

- For each $\hat{x} \in X$ and any $1 \leq i \leq L$ it determines whether $\hat{x} \in X_i$ holds;

- if $\hat{x} \in X_i$ then it supplies a support function $\sigma_{\hat{x}}^i(x)$ to $Z_i(x)$ at $\hat{x}$, otherwise a cut function $\tau_{\hat{x}}^i(x)$ of $X_i(x)$ at $\hat{x}$ $\forall i$. □

There are essentially two different approaches for building algorithms for problems having the structure above: Algorithms based on *aggregate cuts* and methods which employ *disaggregate cuts*. Mixtures of the above strategies are also possible.

Outer approximation using aggregate cuts

The algorithm is based on the following obvious observation: If $x^* \in \hat{X}$ and $\sigma_{x^*}^i(x)$ is a support function for Z_i, $i = 1, \ldots, L$, then

$$\sigma_{x^*}(x) = \sum_{i=1}^{L} \sigma_{x^*}^i(x) \tag{1.46}$$

is a support function for $Z(x)$.

The master problem can now be formulated as follows:

$$\left.\begin{array}{rl}
\min & f(x) + \eta \\
& \sum_{i=1}^{L} \sigma_{x^*}^i(x) - \eta \leq 0, \quad \forall x^* \in \hat{X} \\
& \tau_{x^*}^i(x) \leq 0, \quad \forall x^* \in X \setminus X_i \\
& \hspace{4cm} i = 1, \ldots, L \\
& x \in X.
\end{array}\right\} \tag{1.47}$$

Problem (1.47) is obviously equivalent to (1.45). Due to separability it is sufficient to consider support functions of the form (1.46). The relaxed master problems will have the following form:

$$\left.\begin{array}{rl}
\min & f(x) + \kappa \cdot \eta \\
& \sigma_{x_l}(x) - \eta \leq 0, \quad l = 1, \ldots, p \\
& \tau_{y_k}(x) \leq 0, \quad k = 1, \ldots, q \\
& x \in X
\end{array}\right\} \tag{1.48}$$

where $x_l \in \hat{X}$, $\sigma_{x_l}(x)$ is computed according to (1.46) with $x^* = x_l$, $l = 1, \ldots, p$; $\exists i$ ($1 \leq i \leq L$) such that $y_k \in X \setminus X_i$, $\tau_{y_k}(x) = \tau_{y_k}^i(x)$, $k = 1, \ldots, q$ and κ plays the same role as in (1.26).

Algorithm 1.11 *(Outer approximation scheme, aggregate cuts)*

Step 0. *Initialize.*   Let p=0, q=0, determine an $\hat{x} \in X$ and set $\hat{\eta} = -\infty$. Choose
   a stopping tolerance $\epsilon^* > 0$.

Step 1. *Solve subproblems.*   In turn for $i = 1, \ldots, L$ do: Apply Algorithm 1.10
   to $\hat{x}$ and block i.

- If $\hat{x} \in X_i$ then a support function $\sigma^i_{\hat{x}}(x)$ has been delivered.
- If $\hat{x} \in X \setminus X_i$ then a cut function $\tau^i_{\hat{x}}(x)$ results. In this case set $\tau_{\hat{x}}(x) = \tau^i_{\hat{x}}(x)$ jump out from the loop and **Goto Step 2**.

   At this point for each $i = 1, \ldots, L$ a support function has been computed. Set

$$\sigma_{\hat{x}}(x) = \sum_{i=1}^{L} \sigma^i_{\hat{x}}(x)$$

   and **Goto Step 3**.

Step 2. *Add a feasibility cut.*   Set $q := q + 1$; $y_q := \hat{x}$ and append $\tau_{y_q}(x) \leq 0$ to
   the constraint set of the relaxed master, **Goto Step 5**.

Step 3. *Test for optimality.*   Check whether $\hat{\eta} \geq \sigma_{\hat{x}}(\hat{x}) - \epsilon^*$ holds. If yes then $\hat{x}$
   is ϵ^*-optimal for (1.45) $\Longrightarrow$ **STOP**, otherwise:

Step 4. *Add an optimality cut.*   Set $p := p+1$; $x_p := \hat{x}$ and append $\sigma_{x_p}(x) - \eta \leq 0$
   to the constraint set of the relaxed master.

Step 5. *Solve the new master problem.*   Solve the current relaxed master problem. Let the optimal solution be $(\hat{x}, \hat{\eta})$ with the following convention: As long
   as the relaxed master only contains feasibility cuts, return $\hat{\eta} = -\infty$. **Goto
   Step 1**.   □

All propositions in Section 1.2.1 remain obviously valid as they stand there for the method
outlined above.

Outer approximation with disaggregate cuts

The method is based on the following master problem:

$$\left.\begin{aligned}
\min \mathrm{f}(x) \; + \; &\sum_{i=1}^{L} \eta_i \\
\sigma^i_{x^*}(x) \; - \quad &\eta_i \leq 0, \quad \forall x^* \in X_i, \; i = 1 \ldots, L \\
\tau^i_{x^*}(x) \qquad &\leq 0, \quad \forall x^* \in X \setminus X_i, \; i = 1, \ldots, L \\
x \qquad\quad &\in X.
\end{aligned}\right\} \tag{1.49}$$

Problem (1.49) is obviously equivalent to the original problem (1.45). The relaxed master problems will be the following:

$$
\left.
\begin{aligned}
\min \mathrm{f}(x) \ &+ \sum_{i=1}^{L} \kappa_i \cdot \eta_i \\
\sigma^i_{x_l}(x) \ - \ \ &\quad \eta_i \leq 0, \quad x_l \in X_i, \quad l = 1, \ldots, p \\
&\qquad\qquad\qquad\qquad i \in \{1, \ldots, L\} \\
\tau^i_{y_k}(x) \quad &\qquad \leq 0, \quad y_k \in X \setminus X_i, \quad k = 1, \ldots, q \\
&\qquad\qquad\qquad\qquad i \in \{1, \ldots, L\} \\
x \quad\quad &\qquad \in X
\end{aligned}
\right\}
\qquad (1.50)
$$

where κ_i plays the same role as κ in (1.26).

Algorithm 1.12 *(Outer approximation scheme, disaggregate cuts)*

Step 0. *Initialize.* Let p=0, q=0, determine an $\hat{x} \in X$; set $\hat{\eta}_i = -\infty$, $i = 1, \ldots, L$ and $\hat{\eta} = -\infty$. ($\hat{\eta}$ will be computed as $\hat{\eta} = \sum_{i=1}^{L} \hat{\eta}_i$ in the algorithm). Choose a stopping tolerance $\epsilon^* > 0$.

Step 1. *Solve subproblems.* For $i = 1, \ldots, L$ in turn do: Apply Algorithm 1.10 to $\hat{x}$ and i.

- If $\hat{x} \in X_i$ then a support function $\sigma^i_{\hat{x}}(x)$ has been computed.
- If $\hat{x} \in X \setminus X_i$ then a cut function $\tau^i_{\hat{x}}(x)$ is delivered.

Step 2. *Add feasibility cuts.* For $i = 1, \ldots, L$ in turn do: If $\hat{x} \in X \setminus X_i$ then set $q := q + 1$; $y_q := \hat{x}$ and append $\tau^i_{y_q}(x) \leq 0$ to the constraint set of the relaxed master. If $\hat{x} \in X \setminus X_i$ $\forall i$, then **Goto Step 5**.

Step 3. *Test for optimality.* If $\hat{x} \in X_i$, $\forall i$ then check whether $\hat{\eta} \geq \sum_{i=1}^{L} \sigma^i_{\hat{x}}(\hat{x}) - \epsilon^*$ holds. If yes then $\hat{x}$ is ϵ^*-optimal for (1.45) $\Longrightarrow$ **STOP**, otherwise:

Step 4. *Add optimality cuts.* For $i = 1, \ldots, L$ in turn do: If $\hat{x} \in X_i$ and $\hat{\eta}_i < \sigma^i_{\hat{x}}(\hat{x})$ then set $p := p+1$; $x_p := \hat{x}$ and append $\sigma^i_{x_p}(x) - \eta_i \leq 0$ to the constraint set of the relaxed master.

Step 5. *Solve the new master problem.* Solve the current relaxed master problem. Let the optimal solution be $(\hat{x}, \hat{\eta}_1, \ldots, \hat{\eta}_L)$ with the following convention: As long as the relaxed master does not contain an optimality cut for i, return $\hat{\eta}_i = -\infty$ $\forall i$. Set $\hat{\eta} = \sum_{i=1}^{L} \hat{\eta}_i$ and **Goto Step 1**. $\square$

Again, all propositions in Section 1.2.1 remain valid for the method specified above.

Both algorithms outlined above can be specialized in an obvious way for obtaining the generalized as well as the original Benders decomposition method for dual block-angular problems. It is easy to see that all propositions remain valid with straightforward changes in the assumptions. We start with the nonlinear case. Let us consider the following problem:

$$\left.\begin{array}{ll} \min\left[\, f\left(x\right) + \sum\limits_{i=1}^{L} q_i\left(y^i\right)\right] & \\[2mm] F\left(x\right) \leq b & \\[2mm] T_i\left(x\right) + \quad G\left(y^i\right) \leq h_i, \quad i = 1,\ldots,L & \\[2mm] x \in X & \\[2mm] y^i \in Y_i, \ i = 1,\ldots,L & \end{array}\right\} \qquad (1.51)$$

where

$f : {\rm I\!R}^{n_1} \mapsto {\rm I\!R}^1;\ q_i : {\rm I\!R}^{n_2} \mapsto {\rm I\!R}^1,\ F : {\rm I\!R}^{n_1} \mapsto {\rm I\!R}^{m_1};\ T_i : {\rm I\!R}^{n_1} \mapsto {\rm I\!R}^{m_2};\ G : {\rm I\!R}^{n_2} \mapsto {\rm I\!R}^{m_2},\ \forall i;\ X \subseteq {\rm I\!R}^{n_1},\ Y_i \subseteq {\rm I\!R}^{n_2}\ \forall i.$
Our assumptions are: $f,\ q_i,\ F,\ G,\ T_i,\ i = 1,\ldots,L$, are convex functions F,T_i and G componentwise). $X \neq \emptyset,\ Y_i \neq \emptyset,\ \forall i$; all sets are convex and compact. We assume also that Problem (1.51) has a solution.

Let
$$X = \{\, x \mid x \in X,\ F\left(x\right) \leq b \,\}$$
and
$$X_i = \{\, x \mid x \in X \text{ and } \exists y : G(y) \leq h_i - T_i(x),\ y \in Y_i \,\}.$$
Let $Z_i : X_i \mapsto {\rm I\!R}^1$ defined by

$$\left.\begin{array}{l} Z_i(x) = \min q_i\left(y\right) \\[2mm] \qquad G\left(y\right) \leq h_i - T_i\left(x\right) \\[2mm] \qquad y \in Y_i. \end{array}\right\} \qquad (1.52)$$

We assume regularity: Problem (1.52) is Slater regular $\forall x \in X_i;\ \forall i$.

Our problem can now in an obvious way be reformulated thus taking the form (1.45) above. This implies that both schemes are applicable provided that we specify the sub-algorithm 1.10. This can be done like in Section 1.2.3:

Algorithm 1.13 *(Algorithm for computing support and cut functions)*

Apply a dual adequate convex programming algorithm for (1.52) with $x = \hat{x}$. If $x = \hat{x} \in X_i$ then compute a support function $\sigma_{\hat{x}}^i\left(x\right)$ according to (1.34), otherwise compute a cut function $\tau_{\hat{x}}^i\left(x\right)$ according to (1.35). $\qquad\square$

This way we arrive at the following two algorithms:

Algorithm 1.14 *(Generalized Benders decomposition, aggregate cuts)*

The algorithm is defined by the scheme 1.11 with the sub-algorithm 1.10 replaced by Algorithm 1.13. □

Algorithm 1.15 *(Generalized Benders decomposition, disaggregate cuts)*

The algorithm is defined by the scheme 1.12 with the sub-algorithm 1.10 replaced by Algorithm 1.13. □

Let us finally turn our attention to the linear case. We consider the following problem:

$$\left.\begin{array}{l} \min\,[\,c^T x \,+\, p_1(q^1)^T y^1 \,+\, \ldots \,+\, p_L(q^L)^T y^L\,] \\[4pt] \quad A\,x \hspace{11.5em} = b \\[4pt] \quad T^1 x \,+\, \quad W\,y^1 \hspace{7em} = h^1 \\[4pt] \qquad \vdots \hspace{4em} \ddots \hspace{6.5em} \vdots \\[4pt] \quad T^L x \,+\, \hspace{5em} W\,y^L \quad = h^L \\[4pt] \qquad x \hspace{11em} \geq 0 \\[4pt] \hspace{9em} y^i \quad\;\; \geq 0,\;\; \forall i \end{array}\right\} \qquad (1.53)$$

where the array dimensions correspond to the dimensions in Problem 1.51.

Set $\epsilon^* = 0$ in Algorithms 1.11 and 1.12. To derive the decomposition methods it is sufficient to define the sub-algorithm. This can obviously be done as:

Algorithm 1.16 *(Algorithm for computing support and cut functions)*

Apply the simplex method to the "diagonal" LP problems in turn, with $x = \hat{x}$ fixed. If $x = \hat{x} \in X_i$ then compute a support function $\sigma_{\hat{x}}^i(x)$ according to (1.42), otherwise compute a cut function $\tau_{\hat{x}}^i(x)$ according to (1.43). □

We end up with the following two algorithms:

Algorithm 1.17 *(L-shaped method of Van Slyke and Wets [156])*

The algorithm is defined by the scheme 1.11 with the sub-algorithm 1.10 replaced by Algorithm 1.16. □

Algorithm 1.18 *(Multicut method of Birge and Louveaux, [11])*

The algorithm is defined by the scheme 1.12 with the sub-algorithm 1.10 replaced by Algorithm 1.16.   □

Let us notice that the first appearance of using disaggregate cuts for solving Problem 1.53 by methods rooted in Benders decomposition is in Ruszczyński [138].

Convergence behavior

It is easy to see that all methods in this section have the same convergence behavior as their counterparts with L=1, discussed in the previous sections.

1.3   Regularized outer approximation methods

This section contains the regularized versions of the algorithms in the previous section, except for the cutting plane methods. All algorithms here have their origins in the regularized Benders decomposition method of Ruszczyński [138]. The algorithms presented in this section include Ruszczyński's regularized decomposition method [138] as well as its generalization to the nonlinear case.

In the first subsection a regularized version of the Wolsey scheme is presented. A regularized generalized Benders decomposition method is the subject of the next subsection followed by a subsection devoted to Ruszczyński's regularized Benders decomposition method. The final subsection deals with the special case of dual block-angular problems.

1.3.1   A regularized outer approximation scheme

In this section we consider the same problem (1.21) as in Section 1.2.1, with additional convexity assumptions. Our additional assumptions are: $X_1 \neq \emptyset$; X and X_1 are compact convex sets. Function f is convex on an open set containing X; Z is convex and continuous with respect to X_1. The support functions $\sigma_{x^*}(x)$ and cut functions $\tau_{x^*}(x)$ exist $\forall x^* \in X_1$ and $\forall x^* \in X \setminus X_1$, respectively and they are convex functions.

We also need a regularity condition: $\operatorname{ri} X \cap \operatorname{ri} X_1 \neq \emptyset$.

Let us consider the following regularized relaxed master problem:

$$\left. \begin{array}{ll} \min \; [\, \alpha \cdot \| \, x - z^r \, \|^2 \;\; + \mathrm{f}\,(x) \;\; + Z^p\,(x)\,] & \\[2mm] \qquad\qquad \tau_{y_k}(x) \qquad\qquad\qquad\qquad \leq 0, \quad k = 1, \ldots, q, \\[2mm] \qquad\qquad\quad x \qquad\qquad\qquad\qquad\qquad \in X \end{array} \right\} \qquad (1.54)$$

where $y_k \in X \setminus X_1$, $k = 1, \ldots, q$ and $Z^p(x)$ is the function (1.24).

$z^r \in X_1$ will be called a candidate solution.   Let ζ^r be the "true" objective value corresponding to the candidate, i.e.

$$\zeta^r = f(z^r) + Z(z^r).$$

Parameter α is kept equal to 0 as long as no candidate has been specified; afterwards it is set to 0.5. The specific value 0.5 plays no significant role, it can be replaced by any positive value. It may even vary from iteration to iteration within a finite closed subinterval, not containing 0, of the positive real line, thus playing the role of a control parameter. We will discuss the algorithm with $\alpha = 0.5$ fixed for the sake of simplicity of presentation.

An obviously equivalent form is the following:

$$\left.\begin{array}{rll} \min \alpha \cdot \| x - z^r \|^2 & + f(x) + \eta & \\ \sigma_{x_l}(x) & - \eta \leq 0, & l = 1, \ldots, p \\ \tau_{y_k}(x) & \leq 0, & k = 1, \ldots, q \\ x & \in X & \end{array}\right\} \quad (1.55)$$

where $x_l \in X_1$, $l = 1, \ldots, p$ and $y_k \in X \setminus X_1$, $k = 1, \ldots, q$.

Problem (1.55) is called the regularized relaxed master problem. Let us introduce the following notation:

$$Q^r(x, \eta) = \alpha \cdot \| x - z^r \|^2 + f(x) + \eta$$

It is clear that the solution of Problem (1.55) is unique if $\alpha > 0$ holds.

Proposition 1.12 *Let $(\hat{x}, \hat{\eta})$ be an optimal solution of (1.55) with the property $\hat{x} \in X_1$ then the following inequalities hold:*

$$f(\hat{x}) + \hat{\eta} \leq \alpha \cdot \| \hat{x} - z^r \|^2 + f(\hat{x}) + \hat{\eta} \leq f(z^r) + Z(z^r) = \zeta^r \quad (1.56)$$

$$\hat{\eta} \leq Z(\hat{x}) \quad (1.57)$$

$$\theta^* \leq \zeta^r \quad (1.58)$$

$$\theta^* \leq f(\hat{x}) + Z(\hat{x}). \quad (1.59)$$

Proof: The inequalities (1.56) follow by substituting $(z^r, Z(z^r))$ into the objective of (1.55). The rest of the inequalities directly follow from the assumption or from the definiton of the quantities involved.   □

Observe that the inequalities above do not give a bracket on the optimal objective value θ^* of (1.20), as the analogue inequalities (1.27) in the Wolsey decomposition do. Lower bounds are missing. Lower bounds could be derived in the case $\hat{x} \in X_1$ by solving (1.55) without the regularizing term.

A regularized outer approximation algorithm is formulated below:

Algorithm 1.19 *(Regularized outer approximation scheme)*

Step 0. *Initialize.* Let p=0, q=0, r=0, $\alpha = 0$ determine an $\hat{x} \in X$ and set $\hat{\eta} = \infty$. Choose a starting value $\epsilon_0 > 0$ of the tolerance for candidate changes.

Step 1. *Solve a subproblem.* Apply Algorithm 1.0 to $\hat{x}$. If $\hat{x} \in X_1$ then a support function $\sigma_{\hat{x}}(x)$ is delivered, **Goto Step 3**, otherwise a cut function $\tau_{\hat{x}}(x)$ is available.

Step 2. *Add a feasibility cut.* Set $q := q + 1$; $y_q := \hat{x}$ and append $\tau_{y_q}(x) \leq 0$ to the constraint set of the relaxed master, **Goto Step 5**.

Step 3. *Test for changing the candidate.* Check whether $\hat{\eta} \geq \sigma_{\hat{x}}(\hat{x}) - \epsilon_r$ holds.

- If yes then change the candidate: Set $\epsilon_{r+1} := 0.5\,\epsilon_r$; $r := r + 1$; $z^r := \hat{x}$; $\zeta^r := f(\hat{x}) + \sigma_{\hat{x}}(\hat{x})$. If $r = 1$ then set $\alpha := 0.5$ and **Goto Step 4** else **Goto Step 5**.

- Otherwise **Goto next Step**.

Step 4. *Add an optimality cut.* Set $p := p+1$; $x_p := \hat{x}$ and append $\sigma_{x_p}(x) - \eta \leq 0$ to the constraint set of the relaxed master.

Step 5. *Solve the new master problem.* Solve the current regularized relaxed master problem. Let the optimal solution be $(\hat{x}, \hat{\eta})$ with the following convention: As long as the relaxed master only contains feasibility cuts, return $\hat{\eta} = \infty$.

Step 6. Test for optimality If a candidate already exists, i.e. $r > 0$ and $f(\hat{x}) + \hat{\eta} \geq f(z^r) + Z(z^r) = \zeta^r$ holds then $\hat{x}$ is optimal for (1.21) $\Longrightarrow$ **STOP**, otherwise **Goto Step 1**.   □

The general idea behind the regularization is the following: For the sake of stability a feasible solution, the candidate z^r, is kept as long as the approximation of the objective is not "good enough" in the vicinity of it.

Proposition 1.13

1. *None of the cut functions can reappear, i.e.* $\tau_{\hat{x}}(x) \not\equiv \tau_{y_k}(x)$, *on* X, $k = 1, \ldots, q$.

2. *If a support function reappears, i.e.* $\exists k\ (1 \leq k \leq p) : \sigma_{\hat{x}}(x) \equiv \sigma_{x_k}(x)$, *then a candidate change occurs.*

Proof:

(1) Assume that $\exists k : (1 \leq k \leq q)$ such that $\tau_{\hat{x}}(x) \equiv \tau_{y_k}(x)$. On the one hand we have $\tau_{\hat{x}}(\hat{x}) > 0$ because of the definition of a cut function, and on the other hand $\tau_{y_k}(\hat{x}) \leq 0$ because $\hat{x}$ is a feasible solution of the current relaxed master. This is a contradiction.

(2) As $(\hat{x}, \hat{\eta})$ is a feasible solution of the relaxed master, it follows $\sigma_{x_k}(\hat{x}) \leq \hat{\eta}$ which implies $\sigma_{\hat{x}}(\hat{x}) \leq \hat{\eta}$, i.e. the criterion for changing the candidate is fulfilled. $\square$

Let us denote $(\hat{x}, \hat{\eta})$ at iteration k by (x^k, η^k).

Proposition 1.14 *Let us assume that $r > 0$ holds. Then after performing either Step 2 or Step 4 the optimal objective value of the regularized relaxed master strictly increases.*

Proof: Both type of cuts cut off the previous optimal solution and thus the assertion follows from the uniqueness of the solution of (1.55). $\square$

Proposition 1.15 *For the objective values at the candidates the following inequality holds:*

$$\zeta^{r+1} \leq \zeta^r + \epsilon_r. \tag{1.60}$$

Proof: The criterion for the candidate change with $z^{r+1} = \hat{x}$ yields

$$\begin{aligned}
\zeta^{r+1} &= f(z^{r+1}) + Z(z^{r+1}) = f(\hat{x}) + Z(\hat{x}) = \\
&\quad f(\hat{x}) + \sigma_{\hat{x}}(\hat{x}) \leq f(\hat{x}) + \hat{\eta} + \epsilon_r
\end{aligned}$$

and the proposition follows from (1.56) $\square$

Proposition 1.16 *Let us assume that at Step 3 a candidate change occurs and $(\hat{x}, \hat{\eta})$ solves also (1.55) without the regularizing terms. Then $z^{r+1} = \hat{x}$ is an ϵ_r solution of problem (1.21).*

Proof: The assertion follows immediately from the criterion for a candidate change:

$$\zeta^{r+1} = f(\hat{x}) + Z(\hat{x}) \leq f(\hat{x}) + \hat{\eta} + \epsilon_r \leq f(x^*) + \eta^* + \epsilon_r$$

$\square$

Notice that in the case of an infinite sequence of candidates

$$\sum_{r=1}^{\infty} \epsilon_r = \epsilon_0 < +\infty \tag{1.61}$$

holds.

In what follows we will need Theorem 27.4 of Rockafellar [136]. For the sake of easy reference we formulate a special case as needed in the sequel:

Proposition 1.17 *Rockafellar [136]*
Let $f(x)$ *be a proper convex function and* C *a nonempty convex set. Let us assume that* $\mathrm{ri}(\mathrm{dom}\, f) \cap \mathrm{ri}\, C \neq \emptyset$ *holds. Then* $x^* \in C$ *is a minimizer of* f *relative to* $C \Longleftrightarrow$ $\exists \xi \in \partial f(x^*)$ *such that* $\xi^{\mathrm{T}}(y - x^*) \geq 0 \ \forall y \in C.$

Proof: See Rockafellar [136].   □

Convergence behavior

Proposition 1.18 *The algorithm does not cycle.*

Proof: Obviously a feasibility cut cuts off $\hat{x}$ whereas an optimality cut cuts off $(\hat{x}, \hat{\eta})$. □

Notice that $\hat{x}$ may reappear if Step 4 was executed. In this case in view of Proposition 1.13 the candidate changes.

Proposition 1.19 *If at Step 6 the algorithm stops, then* $\hat{x}$ *is an optimal solution of* *(1.21).*

Proof: In our case inequality (1.56) implies that $\hat{x} = z^r$ holds. Thus $\hat{x} \in X_1$ is fulfilled. Observe that the subgradients of the objective at $(z^r, \hat{\eta})$ are the same for the regularized relaxed master (1.55) with- and without the regularizing term. This implies that fulfillment of any optimality criterion based on subgradients of the objective occurs simultaneously to both problems at this specific point. Because of the regularity condition it follows from Proposition 1.17 that $(\hat{x}, \hat{\eta})$ is an optimal solution of the relaxed master (1.26) without the regularizing term. This fact, the stopping criterion and (1.58) imply the assertion.   □

Proposition 1.20 *If the algorithm generates an infinite sequence of candidates then this sequence converges to a solution of (1.21).*

Proof: The proof is based on a combination of ideas in Rockafellar [137] and Ruszczyński [138]. Let (x^*, η^*) be an optimal solution of (1.23). Let r be fixed and consider the iteration where the next candidate is generated; i.e. $z^{r+1} = \hat{x}$.

The point (x^*, η^*) is feasible for the regularized relaxed master problem (1.55), and $(z^{r+1}, \hat{\eta})$ is the optimal solution. As $(x^* - z^{r+1}, \eta^* - \hat{\eta})$ is a feasible direction at $(z^{r+1}, \hat{\eta})$, from Proposition 1.17 it follows that $\exists \xi \in \partial f(z^{r+1})$ such that the following inequality is fulfilled (remind that $\alpha = 0.5$ holds):

$$(z^{r+1} - z^r)^{\mathrm{T}}(x^* - z^{r+1}) + \xi^{\mathrm{T}}(x^* - z^{r+1}) + \eta^* - \hat{\eta} \geq 0. \tag{1.62}$$

Utilizing the convexity of f results in:

$$(z^{r+1} - z^r)^{\mathrm{T}}(z^{r+1} - x^*) \leq f(x^*) + \eta^* - (f(z^{r+1}) + \hat{\eta}). \tag{1.63}$$

An obvious reformulation yields

$$\| z^{r+1} - x^* \|^2 - \| z^r - x^* \|^2 + (x^* - z^r)^{\mathrm{T}}(z^{r+1} - z^r) \leq f(x^*) + \eta^* - (f(z^{r+1}) + \hat\eta)$$

which can further be reformulated as

$$\| z^{r+1} - x^* \|^2 - \| z^r - x^* \|^2 + \| z^{r+1} - z^r \|^2 - \\ (z^{r+1} - x^*)^{\mathrm{T}}(z^{r+1} - z^r) \leq f(x^*) + \eta^* - (f(z^{r+1}) + \hat\eta).$$

Deleting the term $\| z^{r+1} - z^r \|^2$ and utilizing (1.63) leads to

$$\| z^{r+1} - x^* \|^2 - \| z^r - x^* \|^2 \leq 2\left[(f(x^*) + \eta^*) - (f(z^{r+1}) + \hat\eta)\right]. \tag{1.64}$$

On the other hand, the candidate-change criterion yields:

$$\begin{aligned} f(z^{r+1}) + \hat\eta &\geq f(z^{r+1}) + \sigma_{\hat x}(\hat x) - \epsilon_r = f(z^{r+1}) + Z(z^{r+1}) - \epsilon_r \\ &\geq f(x^*) + \eta^* - \epsilon_r. \end{aligned} \tag{1.65}$$

Combining this inequality with (1.64) results in:

$$\| z^{r+1} - x^* \|^2 - \| z^r - x^* \|^2 \leq 2\,\epsilon_r.$$

As an immediate consequence of this inequality and (1.61) we have:

$$\lim_{r \to \infty} \| z^r - x^* \| = \mu < \infty \tag{1.66}$$

holds. Let us utilize the candidate-change criterion again. After an obvious reformulation of this criterion we get:

$$f(\hat x) + \hat\eta \geq \zeta^{r+1} - \epsilon_r.$$

Substituting into (1.64) results in:

$$\| z^{r+1} - x^* \|^2 - \| z^r - x^* \|^2 \leq 2\left[f(x^*) + \eta^* - \zeta^{r+1}\right] + 2\epsilon_r.$$

Reformulating this inequality yields:

$$\zeta^{r+1} - (f(x^*) + \eta^*) \leq \frac{1}{2}[\| z^r - x^* \|^2 - \| z^{r+1} - x^* \|^2 + 2\epsilon_r].$$

If we now utilize the limes relations (1.66) and (1.61) we get:

$$\sum_{r=1}^{\infty} [\zeta^{r+1} - (f(x^*) + \eta^*)] < \infty$$

which in particular implies:

$$\lim_{r \to \infty} \zeta^r = f(x^*) + \eta^*$$

i.e. the objective values converge along the sequence of candidate solutions to the optimal objective value. Let $\tilde z$ be an accumulation point of the sequence z^r, $r = 1, \ldots$, then

$\tilde{z} \in X_1$ holds and $\tilde{z}$ is an optimal solution of our problem. Repeating now our reasoning with $\tilde{z}$ playing the role of x^*, we get the limes relation (1.66) in the following form:

$$\lim_{r \to \infty} \| z^r - \tilde{z} \| = \nu < \infty.$$

But $\tilde{z}$ being the limit point for a subsequence we have convergence to $\tilde{z}$ implying that $\nu = 0$ holds.

□

As for the Wolsey-scheme this algorithm is also well-defined and stopping implies that an optimal solution has been found. The iteration loop of the algorithm contains the following sub-iteration loops: Let us call major iterations those iterations where the candidate is changed. The algorithm may begin with iterations involving only feasibility cuts (feasibility loop). This may be an infinite sequence thus never coming to the stopping criterion. Assume that after a finite number of iterations Step 3 occurs. The major iterations contain a minor iterations loop between two changes of the candidate. For the algorithm to make sense, it must be ensured that the minor iterations loop is finite. This is in general not true, e.g. it may contain an infinite feasibility loop. Assuming however, that an infinite sequence of candidates is generated, it follows that this sequence converges to an optimal solution of our problem.

1.3.2   Regularized generalized Benders decomposition

In this section we consider the same problem (1.30), defined on Page 14, as for the original generalized Benders decomposition discussed in Section 1.2.3, under the assumptions specified there. The regularized version of the generalized Benders-decomposition algorithm fits into the framework given in the previous section, with the support- and cut functions computed as in Section 1.2.3.

Algorithm 1.20 *(Regularized generalized Benders decomposition)*

The algorithm is defined by the regularized outer approximation scheme 1.19 with the sub-algorithm 1.0. on Page 9 replaced by Algorithm 1.6.

□

Convergence behavior

Proposition 1.21 *Assume that the number of subsequent feasibility cuts is finite throughout. Then either the algorithm terminates in a finite number of steps with an optimal solution or an infinite number of candidate solutions will be generated which converge to a solution of (1.30).*

Proof: In view of Proposition 1.20 it is sufficient to prove that the cycle of iterations between two changes of the candidate is finite. Due to our assumption the first candidate is introduced after a finite number of feasibility cuts. Let us assume that at some iteration a change of the candidate occurs and the criterion for changing the candidate is no more fulfilled at subsequent iterations. Due to our assumption this infinite sequence contains an infinite subsequnce where at Step 4 an optimality cut has been generated. Employing the same proof as Geoffrion [49] in Theorem 2.5 this leads to a contradiction as follows. Let us introduce the following denotations: i_k, $k = 1, \ldots$ are the iteration counters in this subsequence; (x^{i_k}, η^{i_k}) is the solution obtained at Step 5 with u^{i_k} being a corresponding dual solution; $\sigma_{i_k}(x)$ is the support function generated.

Under our assumptions X_1 is a closed set, see Hogan [65]. Because of the compactness assumptions we may assume that $(x^{i_k}, \eta^{i_k}) \to (\tilde{x}, \tilde{\eta})$, $(k \to +\infty)$ holds with $\tilde{x} \in X_1$. The Slater condition implies that the set of dual solutions is uniformly bounded in a neighbourhood of $\tilde{x}$, see Geoffrion [49]. So we may assume that $u^{i_k} \to \tilde{u}$ $(k \to +\infty)$ also holds. Because at iteration i_k we added the constraint $\gamma(x, u^{i_k}) \le \eta$ to the regularized relaxed master, see Proposition 1.4 and equation (1.34), for the subsequence we clearly have

$$\eta^{i_{k+1}} \ge \gamma(x^{i_{k+1}}, u^{i_k})$$

which implies by the continuity of γ, see Meyer [113]

$$\tilde{\eta} \ge \gamma(\tilde{x}, \tilde{u}). \tag{1.67}$$

Next we observe that $\tilde{u}$ is an optimal dual solution for (1.10) on Page 4, with $x = \tilde{x}$ fixed. In fact this property follows from the upper semicontinuity of the point to set map $D(x) = \{ u \mid \gamma(x, u) = Z(x), u \ge 0 \}$, see Geoffrion [49]. Using this fact and the strong duality theorem we get from the previous inequality

$$\tilde{\eta} \ge Z(\tilde{x}). \tag{1.68}$$

On the other hand in Step 4 of the algorithm we have

$$\eta^{i_k} < Z(x^{i_k}) - \epsilon_r$$

so using the continuity of f and Z (see Meyer [113]) we get in the limit

$$\tilde{\eta} \le Z(\tilde{x}) - \epsilon_r \tag{1.69}$$

which clearly contradicts (1.68).   □

Concerning our assumption on feasibility cuts we mention that it is obviously fulfilled if the second stage is linear or if $X_1 = X$ holds (complete recourse). In the general case the same considerations apply as in Geoffrion [49], or Flippo and Rinnooy Kan [42].

1.3.3   The regularized decomposition method of Ruszczyński

We consider the linear problem

$$
\left.
\begin{aligned}
\min [\, \mathrm{c}^{\mathrm{T}} x + \mathrm{q}^{\mathrm{T}} y \,] & \\
\mathrm{A}\, x \qquad\quad &= \mathrm{b} \\
\mathrm{T}\, x + \mathrm{W}\, y \ &= \mathrm{h} \\
x \qquad\quad &\geq 0 \\
y \ &\geq 0.
\end{aligned}
\right\}
\qquad (1.70)
$$

First we discuss the decomposition method without cut-dropping in the regularized relaxed master.

Algorithm 1.21 *(Regularized decomposition method of Ruszczyński; version without cut dropping)*

The algorithm is defined by the regularized outer approximation scheme 1.19 with the following modifications: The sub-algorithm 1.0 is replaced by Algorithm 1.8 and for the candidate-change tolerance is set to $\epsilon_0 = 0$.   □

Notice that $\epsilon_0 = 0$ implies $\epsilon_r = 0$ throughout. It is easy to see that all propositions in Sections 1.3.1 and 1.3.2 remain valid for this case.

Convergence behavior

Proposition 1.22 *The number of subsequent iterations between two changes of the candidate is finite.*

Proof: As the sets of different support- and cut-functions are finite, the assertion follows from Proposition 1.14.   □

Notice that the previous Proposition could have also been derived from Proposition 1.13. In view of this Proposition it is clear that in the version without cut dropping the number of feasibility cuts, where $\hat{x} \in X_1$ is not fulfilled, is finite. In other words the first candidate appears after a finite number of feasibility cuts. The previous proof has been chosen for the reason that it remains valid also then, when slack cuts are dropped from the relaxed regularized master.

The key to the finiteness of the procedure is the following theorem of Ruszczyński, see [138]:

Proposition 1.23 *The number of iterations involving candidate changes, where the new candidate does not also solve the relaxed master (1.26) without the regularizing term in the objective, is finite.*

Proof: Let us assume that the algorithm generates an infinite sequence of candidates. We know from Proposition 1.20 that in this case $\lim_{r \to \infty} z^r = x^*$ holds with x^* being a solution of (1.70). Let us also assume that there exists a subsequence $\{r_k\}_{k=1}^{\infty}$ such that $(z^{r_k+1}, \hat{\eta})$ is not a solution of (1.23) on Page 8.

Let d_k be the negative gradient of the objective in the regularized relaxed master (1.55) (remind that $\alpha = 0.5$ holds) i.e.

$$d_k = -\nabla Q^{r_k}(z^{r_k+1}, \hat{\eta}) = -(z^{r_k+1} - z^{r_k} + c, 1)^{\mathrm{T}}$$

and d the negative gradient of the objective in the relaxed master (1.26) without regularization,

$$d = -(c, 1)^{\mathrm{T}}.$$

Observe that the following limit relation holds:

$$\lim_{k \to \infty} d_k = d.$$

Notice that the current iterate $(\hat{x}, \hat{\eta})$ with $\hat{x} = z^{r_k+1}$ solves the regularized relaxed master (1.55), for each k. This implies in view of Proposition 1.17 that d_k belongs to the normal cone of the feasible set at $(\hat{x}, \hat{\eta})$. The number of possible different feasible sets in (1.55) is finite each of them having a finite number of possible different normal cones. This implies that there exists a fixed normal cone C such that $d_k \in C$ holds for an infinite subsequence of r_k. The cone C being closed we get that $d \in C$ also holds thus implying by Proposition 1.17 that for the above-mentioned subsequence $(\hat{x}, \hat{\eta})$ is a solution of the relaxed master (1.55) without regularization. This is a contradiction. $\square$

Proposition 1.24 *The algorithm terminates in a finite number of iterations at an optimal solution of (1.70).*

Proof: The algorithm either stops at Step 6 with an optimal solution, or as a consequence of Propositions 1.22 and 1.23 after a finite number of iterations we arrive at an iteration with candidate-change, where $(\hat{x}, \hat{\eta})$ also solves the relaxed master without the regularizing term. According to Proposition 1.16 this implies that $z^{r+1} = \hat{x}$ is an optimal solution of our problem (1.21) on Page 7, i.e $\zeta^{r+1} = \theta^*$ holds with θ^* being the optimal objective value of (1.21). According to Proposition 1.22 after a finite number of iterations we obtain at Step 5 a solution which fulfills the candidate-change criterion, i.e. $f(\hat{x}) + \hat{\eta} \geq f(\hat{x}) + F(\hat{x}) \geq \theta^* = \zeta^{r+1}$ holds. This means that the optimality criterion at Step 6 is fulfilled and the algorithm stops. $\square$

Now we present the full algorithm of Ruszczyński [138]. There are two additional features in comparison with Algorithm 1.21. On the one hand a cut dropping strategy is implemented and on the other hand a heuristic rule added for changing the candidate.

Considering cut dropping, the single point where difficulties may arise is the Phase I procedure consisting in finding the first candidate. Otherwise cuts may freely be dropped in (1.55) as long as the solution of the regularized relaxed master remains the same.

The current set of constraints in the regularized relaxed master (1.55) will be called like in [138] *a committee.*

For a cut-dropping strategy Ruszczyński [138] suggests to keep the minimum amount of cuts leaving the solution of (1.55) unchanged: Keep the cuts having strictly positive Lagrange multipliers at the solution of (1.55). According to the discussion after Proposition 1.3 there exist Lagrange multipliers such that the gradients of constraints corresponding to nonzero multipliers are linearly independent. Use an algorithm for solving (1.55) which terminates with a solution having the property that the set of constraints corresponding to positive Lagrange multipliers contains linearly independent gradient vectors. The rest of constraints can be dropped without changing the solution. This implies that at most $n+1$ constraints remain.

There are at least two reasonable possibilities for finding a starting $x^0 \in X_1$. One possibility is not to drop cuts as long as no candidate has been found. The other way is to apply the algorithm below with cut dropping, to an auxiliary first-phase problem with artificial variables (this method is employed in [138]).

Algorithm 1.22 *(Regularized decomposition method of Ruszczyński)*

Step 0. *Initialize.*   Let p=0, q=0, r=0, $\alpha = 0$ determine an $\hat{x} \in X_1$ and set $\hat{\eta} = \infty$.

Step 1. *Solve a subproblem.*   Apply the simplex method for solving (1.37) on Page 16, with $x = \hat{x}$. If $x = \hat{x} \in X_1$ then compute a support function $\sigma_{\hat{x}}(x)$ according to (1.39) and **Goto Step 3**, otherwise compute a cut function $\tau_{\hat{x}}(x)$ according to (1.40).

Step 2. *Add a feasibility cut.*   Set $q := q + 1$; $y_q := \hat{x}$ and append $\tau_{y_q}(x) \leq 0$ to the constraint set of the relaxed master, **Goto Step 5**.

Step 3. *Test for changing the candidate.*   Check whether $\hat{\eta} \geq \sigma_{\hat{x}}(\hat{x})$ holds.

- If yes then change the candidate: Set $r := r + 1$; $z^r := \hat{x}$; $\zeta^r := f(\hat{x}) + \sigma_{\hat{x}}(\hat{x})$. If $r = 1$ then set $\alpha := 0.5$ and **Goto Step 4** else **Goto Step 5**.

- Otherwise test heuristic rules for changing the candidate. For these rules it must hold that they are fulfilled only for a finite number of iterations. Ruszczyński [138] suggests the following rule: Let $0 < \gamma < 1$ be fixed. Check whether there are exactly $n+1$ cuts after Step 7 and whether $f(\hat{x}) + Z(\hat{x}) \leq \gamma \zeta^r + (1 - \gamma)(f(\hat{x}) + \hat{\eta})$ holds. (ζ^r was defined as $\zeta^r = f(z^r) + Z(z^r)$).

- If yes then change the candidate: Set $r := r + 1$; $z^r := \hat{x}$; $\zeta^r := f(\hat{x}) + \sigma_{\hat{x}}(\hat{x})$; if $r = 1$ then set $\alpha := 0.5$. (Heuristic candidate change)
- Otherwise no candidate change occurs.

In both cases **Goto next Step**.

Step 4. *Add an optimality cut.* Set $p := p+1$; $x_p := \hat{x}$ and append $\sigma_{x_p}(x) - \eta \leq 0$ to the constraint set of the relaxed master.

Step 5. *Solve the new master problem.* Solve the current regularized relaxed master problem. Use an algorithm which terminates with a solution having the property that the set of constraints corresponding to positive Lagrange multipliers contains linearly independent gradient vectors. Let the optimal solution be $(\hat{x}, \hat{\eta})$.

Step 6. Test for optimality   If $f(\hat{x}) + \hat{\eta} \geq f(z^r) + Z(z^r) = \zeta^r$ holds then $\hat{x}$ is optimal for $(1.21) \Longrightarrow$ **STOP**, otherwise:

Step 7. Drop superfluous cuts   Drop all cuts having zero Lagrange multipliers at the optimal solution of the regularized relaxed master. This implies that at most n+1 constraints remain there; **Goto Step 1**.

$\square$

Notice that in the algorithm as formulated above we started from a feasible (i.e. $\hat{x} \in X_1$) solution. As already mentioned above, the algorithm itself can be used for determining such a point, when applied to a Phase-I problem, see Ruszczyński [138].

Convergence behavior

Proposition 1.25   *The heuristic rule specified at Step 3 for candidate change is only fulfilled at a finite number of iterations.*

Proof: Let us suppose that the algorithm does not stop at Step 6. From the heuristic rule it is clear that its fulfillment implies $\zeta^{r+1} < \zeta^r$. As n+1 independent active cuts uniquely determine the point $(\hat{x}, \hat{\eta})$ it follows that the committees at Steps 3 are different for each case when a heuristic candidate change occurs. The number of different committees being finite the proposition follows. $\square$

Proposition 1.26 *The regularized decomposition method of Ruszczyński terminates in a finite number of iterations at an optimal solution of (1.70).*

Proof: By utilizing the previous Proposition, considering the fact that Proposition 1.22 remains valid under the cut dropping strategy of the algorithm, and taking into account our remarks concerning cut dropping, the finiteness of the procedure can now be proved exactly in the same way as in the case of Proposition 1.24. $\square$

1.3.4 Regularized decomposition methods for dual block–angular problems

The regularized algorithms defined in the previous sections can be specialized for dual block–angular problems in the same way as it has been done in Section 1.2.5 for the algorithms without regularization. It is easy to check that all Propositions in Sections 1.3.1, 1.3.2 and 1.3.3 carry over in a straightforward manner to the corresponding algorithms specified below, which we list for the sake of completeness.

Let us consider Problem (1.45) first. Our assumptions now involve convexity and are as follows: $X_i \neq \emptyset$, $i = 1, \ldots, L$; X is compact and convex; X_i, $i = 1, \ldots, L$ are closed convex sets. The function f is convex on an open set containing X; Z_i is continuous with respect to X_i, $i = 1, \ldots, L$; $Z_i, i = 1, \ldots, L$ are convex functions. For each $x^* \in X_i$ there exists a continuous and convex support function $\sigma_{x^*}^i(x)$ to Z_i at x^*, $i = 1, \ldots, L$ and for each $x^* \in X \setminus X_i$ there exists a continuous and convex cut function $\tau_{x^*}^i(x)$ of X_i at x^*, $i = 1, \ldots, L$. We assume that the regularity condition $\operatorname{ri} X \cap \operatorname{ri} \check{X}_i \neq \emptyset, i = 1, \ldots, L$ holds.

Regularized outer approximation using aggregate cuts

The relaxed master problem (1.48) on Page 19 is now replaced by its regularized counterpart:

$$\left. \begin{array}{lll} \min \alpha \cdot \| x - z^r \|^2 & + \mathrm{f}(x) & + \kappa \cdot \eta \\ \sigma_{x_l}(x) & - \eta \leq 0, & l = 1, \ldots, p \\ \tau_{y_k}(x) & \leq 0, & k = 1, \ldots, q \\ x & \in X \end{array} \right\} \quad (1.71)$$

with z^r being the current candidate; $x_l \in \hat{X}$; $\sigma_{x_l}(x)$ being computed as (1.46) with $x^* = x_l$, $l = 1, \ldots, p$; $\exists i\ (1 \leq i \leq L)$ such that $y_k \in X \setminus X_i$, $\tau_{y_k}(x) = \tau_{y_l}^i(x)$, $k = 1, \ldots, q$.

Algorithm 1.23 *(Regularized outer approximation scheme, aggregate cuts)*

Step 0. *Initialize.* Let p=0, q=0, r=0, α=0; determine an $\hat{x} \in X$ and set $\hat{\eta} = \infty$. Choose a starting value $\epsilon_0 > 0$ for the candidate-change tolerance.

Step 1. *Solve subproblems.* In turn for $i = 1, \ldots, L$ do: Apply Algorithm 1.10 to $\hat{x}$ and block i.

- If $\hat{x} \in X_i$ then a support function $\sigma_{\hat{x}}^i(x)$ has been delivered.

- If $\hat{x} \in X \setminus X_i$ then a cut function $\tau_{\hat{x}}^i(x)$ results. In this case set $\tau_{\hat{x}}(x) = \tau_{\hat{x}}^i(x)$ and jump out from the loop, **Goto Step 2**.

For each $i = 1, \ldots, L$ a support function has been computed. Set

$$\sigma_{\hat{x}}(x) = \sum_{i=1}^{L} \sigma_{\hat{x}}^{i}(x)$$

and **Goto Step 3**.

Step 2. *Add a feasibility cut.* Set $q := q + 1$; $y_q := \hat{x}$ and append $\tau_{y_q}(x) \leq 0$ to the constraint set of the relaxed master, **Goto Step 5**.

Step 3. *Test for changing the candidate.* Check whether $\hat{\eta} \geq \sigma_{\hat{x}}(\hat{x}) - \epsilon_r$ holds.

- If yes then change the candidate: Set $\epsilon_{r+1} := 0.5\,\epsilon_r$; $r := r + 1$; $z^r :=$ $\hat{x}$; $\zeta^r := f(\hat{x}) + \sigma_{\hat{x}}(\hat{x})$. If $r = 1$ then set $\alpha := 0.5$ and **Goto Step 4** else **Goto Step 5**.

- Otherwise **Goto next Step**.

Step 4. *Add an optimality cut.* Set $p := p+1$; $x_p := \hat{x}$ and append $\sigma_{x_p}(x) - \eta \leq 0$ to the constraint set of the relaxed master.

Step 5. *Solve the new master problem.* Solve the current relaxed master problem (1.71). Let the optimal solution be $(\hat{x}, \hat{\eta})$ with the following convention: As long as the relaxed master only contains feasibility cuts, return $\hat{\eta} = \infty$.

Step 6. Test for optimality If a candidate already exists, i.e. $r > 0$ and $f(\hat{x}) + \hat{\eta} \geq$ $f(z^r) + Z(z^r) = \zeta^r$ holds then $\hat{x}$ is optimal for (1.45) $\Longrightarrow$ **STOP**, otherwise **Goto Step 1**. $\qquad\qquad\Box$

Regularized outer approximation with disaggregate cuts

This method is based on the regularized version of (1.50) as formulated below:

$$\left.\begin{aligned}
\min f(x) \; + \; \alpha \cdot \| x - z^r \|^2 \; + \; \sum_{i=1}^{L} \kappa_i \eta_i & \\
\sigma_{x_l}^{i}(x) \; - \qquad\qquad\qquad\quad \eta_i \leq 0, \quad x_l \in X_i, \quad l = 1, \ldots, p & \\
\qquad\qquad\qquad\qquad\qquad\qquad i \in \{1, \ldots, L\} & \\
\tau_{y_l}^{i}(x) \qquad\qquad\qquad\qquad\quad \leq 0, \quad l = 1, \ldots, q & \\
\qquad\qquad\qquad\qquad\qquad\qquad i \in \{1, \ldots, L\} & \\
x \qquad\qquad\qquad\qquad\qquad\quad \in X &
\end{aligned}\right\} \quad (1.72)$$

where z^r is the current candidate; $\exists i \; (1 \leq i \leq L)$ such that $y_l \in X \setminus X_i$, $\tau_{y_l}(x) = \tau_{y_l}^{i}(x)$, $(l = 1, \ldots, q)$.

Algorithm 1.24 *(Regularized outer approximation scheme, disaggregate cuts)*

Step 0. *Initialize.*   Let p=0, q=0, r=0, α=0; determine an $\hat{x} \in X$; set $\hat{\eta}_i = -\infty$, $i = 1, \ldots, L$ and $\hat{\eta} = -\infty$. ($\hat{\eta}$ will be computed as $\hat{\eta} = \sum_{i=1}^{L} \hat{\eta}_i$ in the algorithm). Choose a starting value $\epsilon_0 > 0$ for the candidate-change tolerance.

Step 1. *Solve subproblems.*   For $i = 1, \ldots, L$ in turn do: Apply Algorithm 1.10 to $\hat{x}$ and i.

- If $\hat{x} \in X_i$ then a support function $\sigma_{\hat{x}}^i(x)$ has been computed.
- If $\hat{x} \in X \setminus X_i$ then a cut function $\tau_{\hat{x}}^i(x)$ is delivered.

Step 2. *Add feasibility cuts.*   For $i = 1, \ldots, L$ in turn do: If $\hat{x} \in X \setminus X_i$ then set $q := q + 1$; $y_q := \hat{x}$ and append $\tau_{y_q}^i(x) \leq 0$ to the constraint set of the relaxed master. If $\hat{x} \in X \setminus X_i \; \forall i$, then **Goto Step 5**.

Step 3. *Test for changing the candidate.*   If $\hat{x} \in X_i \; \forall i$ then compute $\sigma_{\hat{x}}(\hat{x})$ according to (1.46) and check whether $\hat{\eta} \geq \sigma_{\hat{x}}(\hat{x}) - \epsilon_r$ holds.

- If yes then change the candidate: Set $\epsilon_{r+1} := 0.5\,\epsilon_r$; $r := r + 1$; $z^r := \hat{x}$; $\zeta^r := f(\hat{x}) + \sigma_{\hat{x}}(\hat{x})$. If $r = 1$ then set $\alpha := 0.5$ and then **Goto Step 4** else **Goto Step 5**.
- Otherwise go to the next step.

Step 4. *Add optimality cuts.*   For $i = 1, \ldots, L$ in turn do: If $\hat{x} \in X_i$ and $\hat{\eta}_i < \sigma_{\hat{x}}^i(\hat{x})$ then set $p := p+1$; $x_p := \hat{x}$ and append $\sigma_{x_p}(x) - \eta_i \leq 0$ to the constraint set of the relaxed master.

Step 5. *Solve the new master problem.*   Solve the current relaxed master problem (1.72). Let the optimal solution be $(\hat{x}, \hat{\eta}_1, \ldots, \hat{\eta}_L)$ with the following convention: As long as the relaxed master does not contain an optimality cut for i, return $\hat{\eta}_i = \infty \; \forall i$. Set $\hat{\eta} = f(\hat{x}) + \sum_{i=1}^{L} \hat{\eta}_i$.

Step 6. Test for optimality   If $\hat{x} \in X_i \; \forall i$ and a candidate already exists, i.e. $r > 0$ and $f(\hat{x}) + \hat{\eta} \geq f(z^r) + Z(z^r) = \zeta^r$ holds then $\hat{x}$ is optimal for (1.45) $\Longrightarrow$ **STOP**, otherwise **Goto Step 1**.

□

As in Section 1.2.5 let us consider Problem (1.51) next, under the same assumptions as specified there.

The following two algorithms can now be defined:

Algorithm 1.25 *(Regularized generalized Benders decomposition, aggregate cuts)*

The algorithm is defined by the scheme 1.23 with the sub-algorithm 1.10 replaced by Algorithm 1.13.

□

Algorithm 1.26 *(Regularized generalized Benders decomposition, disaggregate cuts)*

The algorithm is defined by the scheme 1.24 with the sub-algorithm 1.10 replaced by Algorithm 1.13. $\qquad\qquad\qquad\qquad\qquad\qquad\qquad\qquad\qquad\qquad\qquad\qquad$ $\square$

Let us finally turn our attention to the linear case. We consider Problem 1.53 having a dual block–angular structure. The sub-algorithm needed in the decomposition method is now clearly Algorithm 1.16 which results in two algorithms when used in the outer approximation schemes 1.23 and 1.24. Utilizing linearity and adding a cut dropping strategy results in the regularized decomposition methods of Ruszczyński. Below we specify the algorithm with disaggregate cuts, the method with aggregete cuts can be constructed similarly.

Algorithm 1.27 *(Regularized decomposition method of Ruszczyński, disaggregate cuts)*

Step 0. *Initialize.* Let p=0, q=0, r=0, $\alpha = 0$, $0 < \gamma < 1$, determine an $\hat{x} \in X_1$; set $\hat{\eta}_i = -\infty$, $i = 1, \ldots, L$ and $\hat{\eta} = -\infty$. ($\hat{\eta}$ will be computed as $\hat{\eta} = \sum_{i=1}^{L} \hat{\eta}_i$ in the algorithm).

Step 1. *Solve subproblems.* For $i = 1, \ldots, L$ in turn do: Apply Algorithm 1.16 to $\hat{x}$ and i.

- If $\hat{x} \in X_i$ then a support function $\sigma_{\hat{x}}^i (x)$ has been computed.
- If $\hat{x} \in X \setminus X_i$ then a cut function $\tau_{\hat{x}}^i (x)$ is delivered.

Step 2. *Add feasibility cuts.* For $i = 1, \ldots, L$ in turn do: If $\hat{x} \in X \setminus X_i$ then set $q := q + 1$; $y_q := \hat{x}$ and append $\tau_{y_q}^i (x) \leq 0$ to the constraint set of the relaxed master. If $\hat{x} \in X \setminus X_i$ $\forall i$, then **Goto Step 5**.

Step 3. *Test for changing the candidate.* If $\hat{x} \in X_i$ $\forall i$ then compute $\sigma_{\hat{x}} (\hat{x})$ according to (1.46) on Page 19 and check whether $\hat{\eta} \geq \sigma_{\hat{x}} (\hat{x}) - \epsilon_r$ holds.

- If yes then change the candidate: Set $r := r + 1$; $z^r := \hat{x}$; $\zeta^r := f (\hat{x}) + \sigma_{\hat{x}} (\hat{x})$; if $r = 1$ then set $\alpha := 0.5$.
- Otherwise test the following heuristic rule for changing the candidate: Check whether there are exactly n+1 cuts after Step 7 and whether $f (\hat{x}) + Z (\hat{x}) \leq \gamma \zeta^r + (1 - \gamma) (f (\hat{x}) + \hat{\eta})$ holds.
  - If yes then change the candidate: Set $r := r + 1$; $z^r := \hat{x}$; $\zeta^r := f (\hat{x}) + \sigma_{\hat{x}} (\hat{x})$; if $r = 1$ then set $\alpha := 0.5$. (Heuristic candidate change)
  - Otherwise no candidate change occurs.

  In both cases **Goto Step 4**.

Step 4. *Add optimality cuts.* For $i = 1, \ldots, L$ in turn do: If $\hat{x} \in X_i$ and $\hat{\eta}_i < \sigma_{\hat{x}}^i (\hat{x})$ then set $p := p+1$; $x_p := \hat{x}$ and append $\sigma_{x_p}^i (x) - \eta_i \leq 0$ to the constraint set of the relaxed master.

Step 5. *Solve the new master problem.*   Solve the current relaxed master problem
(1.72). Use an algorithm which terminates with a solution having the property
that the set of constraints corresponding to positive Lagrange-multipliers has
linearly independent gradient vectors. Let the optimal solution be $(\hat{x}, \hat{\eta}_1, \dots, \hat{\eta}_L)$,
set $\hat{\eta} = \sum_{i=1}^{L} \hat{\eta}_i$.

Step 6. Test for optimality   If $\hat{x} \in X_i \forall i$ and $f(\hat{x}) + \hat{\eta} \geq f(z^r) + Z(z^r) = \zeta^r$ holds
then $\hat{x}$ is optimal for (1.45) $\Longrightarrow$ **STOP**, otherwise:

Step 7. Drop superfluous cuts   Drop all cuts having zero Lagrange multipliers at
the optimal solution of the regularized relaxed master. This implies that at
most n+L constraints remain there; **Goto Step 1.**

$\square$

Considering the starting point the same comment applies as in the previous subsection
with L=1.

1.4   Barrier methods

In this section we give a short summary on barrier methods which are also called interior point methods. We restrict the presentation to the logarithmic barrier method.
This method was introduced by Frisch [40], [41] and further developed by Fiacco and
McCormick [34] in the much broader context of sequential unconstrained minimization
techniques (SUMT).

We consider the following problem:

$$\left. \begin{array}{c} \min f(x) \\ G(x) \geq 0 \end{array} \right\} \tag{1.73}$$

where $f : \mathbb{R}^n \mapsto \mathbb{R}^1$; $G : \mathbb{R}^n \mapsto \mathbb{R}^m$; $G(x) = (g_1(x), \dots, g_m(x))^T$.

Let
$$S = \{ x \mid g_i(x) \geq 0, \ \forall i \} \text{ and } S^\circ = \{ x \mid g_i(x) > 0, \ \forall i \}.$$

Our assumptions are the following: The function f is convex whereas the functions
g_i, $i = 1, \dots, m$ are concave. The set of optimal solutions is nonempty and bounded.
We also assume that $S^\circ \neq \emptyset$ holds. In our convex programming case S° coincides with
the topological interior of S which justifies the denotation.

The following function will be called the *logarithmic barrier function* corresponding to
Problem 1.73:
$$\Phi(x) = - \sum_{i=1}^{m} \log g_i(x).$$

We consider the following family of functions, parametrized by $\mu > 0$:

$$\Psi(x, \mu) = f(x) + \mu \cdot \Phi(x).$$

The logarithmic barrier method is based on the following family of auxiliary unconstrained optimization problems:

$$\min_x \Psi(x, \mu). \tag{1.74}$$

For a fixed $\mu > 0$ let $x(\mu) \in \mathrm{argmin}_x\{\ \Psi(x, \mu)\ \}$. The logarithmic barrier method of Fiacco and McCormick [34] is defined as follows:

Algorithm 1.28 *(Logarithmic barrier method, Fiacco-McCormick)*

Step 0. *Initialize.*   Let $k := 1$, $\epsilon > 0$, $x^1 \in S°$, $\mu_1 > 0$.

Step 1. *Solve an auxiliary problem.*   Using x^k as a starting point solve Problem 1.74 with $\mu = \mu_k$, i.e. compute $x^{k+1} := x(\mu_k)$.

Step 2. *Test for optimality.*   Check whether $m \cdot \mu_k < \epsilon$ holds. If yes then x^{k+1} is an ϵ -optimal solution of (1.73) $\implies$ **STOP**, otherwise:

Step 3. *Reduce the barrier parameter.*   Choose $\mu_{k+1} < \mu_k$ in such a way that the sequence $\{\mu_k\}_{k=1}^{\infty}$ converges to zero. Set $k := k + 1$ and **Goto Step 1**.

$\square$

Convergence behavior

The above specified algorithm converges under our assumptions, see e.g. Fiacco and McCormick [34], Kall [72] or Luenberger [98]. For the stopping rule and for a procedure for finding a starting $x^1 \in S°$ see Fiacco and McCormick [34].

Let us mention that the method also works under additional constraints. More precisely, let us consider the following problem:

$$\left.\begin{array}{l} \min f(x) \\ \quad G(x) \geq 0 \\ \quad x \ \in X \end{array}\right\} \tag{1.75}$$

where X is a closed convex set and we have the following additional assumption: $S° \cap X \neq \emptyset$. The auxiliary problem (1.74) has now the following form:

$$\min_x\{\Psi(x, \mu) \mid x \in X\ \}.$$

The covergence of this modified algorithm can be proved in a similar manner as for the original method, see Bazaraa and Shetty [4].

Let us turn our attention to the auxiliary problem (1.74). Due to the logarithmic term the solution of this problem is unique under quite general assumptions, see Fiacco and McCormick [34] and Jarre [69]. Under the same assumptions the following optimization problem has a unique solution:

$$\min_x \Phi(x).$$

The solution, denoted by x^c, is called the *analytic center* of Problem (1.73). This notion was first introduced and studied by Sonnevend [147].

Let us now consider the set of points $x(\mu)$ under assumptions ensuring uniqueness. Under general assumptions this set is a curve, which is called the *central path*, and has the following properties:

$$\lim_{\mu \to \infty} x(\mu) = x^c \text{ and } \lim_{\mu \to 0} x(\mu) = x^*$$

where x^* is an optimal solution of Problem (1.73), see Fiacco and McCormick [34], McLinden [111] and Monteiro and Zhou [114].

Let us emphasize that neither the analytical center nor the central path are geometric constructs. A simple replication of one of the constraints obviously leaves the set of feasible solutions intact but both the analytical center and the central path will change.

Recently much attention has been paid to logarithmic barrier methods, and in general to interior point methods. Many interesting new algorithms have been developed resulting in an enormous amount of publications. We refer to den Hertog [54] and Jarre [69], and to the references therein. We just emphasize here, that in the nonlinear case the algorithms explicitly work with Hessians, being based on Newton's method for tracing the central path.

1.5   Central cutting plane methods

This section is devoted to central cutting plane methods. These methods are not outer approximation methods, since objective cuts, possibly cutting off portions of the feasible domain, are also employed.

We consider a nonlinear programming problem in the following form:

$$\left. \begin{array}{c} \max c^T x \\ g_i(x) \geq 0, \quad i = 1, \dots, m \\ x \in X \end{array} \right\} \tag{1.76}$$

where $X \subseteq \mathbb{R}^n$; $c \in \mathbb{R}^n$; $g_i : X \mapsto \mathbb{R}^1, i = 1, \dots, m$. Let the feasible domain be denoted by S.

We assume that X is a bounded polyhedron; g_i, $i = 1, \ldots, m$ are concave and continuously differentiable and the feasible set is nonempty. We also assume that for Problem (1.76) the Slater-condition holds. Similarly as in Section 1.2.2 on cutting plane methods we suppose that all of the functions g_i are nonlinear, i.e. linear constraints are incorporated into the defining inequalities of X.

The cutting plane methods of Section 1.2.2 generate a sequence of polyhedrons $\mathcal{P}_1, \mathcal{P}_2, \ldots$ each of them containing the feasible set of (1.76), and use $x^k \in$ argmin $\{ c^\mathrm{T} x \mid x \in \mathcal{P}_k \}$ for approximating an optimal solution. Central cutting plane methods also employ a sequence of polyhedrons $\mathcal{P}_1, \mathcal{P}_2, \ldots$ but now it is only ensured that each of these polyhedrons contains the set of optimal solutions. The "centers" of these polyhedrons are used for approximating an optimal solution.

The general scheme of the central cutting plane method is as follows:

Algorithm 1.29 *(Central cutting plane scheme)*

Step 0. *Initialize.* Let $k := 1$, $\underline{f}$ be a lower bound of $c^\mathrm{T} x$ on S; $\mathcal{P}_0 := \{x \mid x \in X,\ c^\mathrm{T} x \geq \underline{f}\}$.

Step 1. *Compute the "center" of $\mathcal{P}_{k-1}$.* Compute x^k, the "center" of $\mathcal{P}_{k-1}$. Depending how a "center" is interpreted and how it is computed, different algorithms may be designed.

Step 2. *Test for optimality.* If an optimality criterion is fulfilled $\Longrightarrow$ **STOP**. The optimality criterion also varies with the algorithms. If $x^k \in S$ then **Goto Step 4**, otherwise **Goto Step 3**

Step 3. *Add a feasibility cut.* Add the Kelley cut (see Section 1.2.2) to the defining inequalities of $\mathcal{P}_{k-1}$, i.e. let $i_0 :\ g_{i_0}(x^k) = \min_i g_i(x^k)$ and $\mathcal{P}_k := \mathcal{P}_{k-1} \cap \{x \mid g_{i_0}(x^k) + \nabla^\mathrm{T} g_{i_0}(x^k)(x - x^k) \geq 0\}$. Set $k := k+1$ and **Goto Step 1**

Step 4. *Add an objective cut.* Add an objective cut to the defining inequalities of $\mathcal{P}_{k-1}$: $\mathcal{P}_k := \mathcal{P}_{k-1} \cap \{x \mid c^\mathrm{T} x - c^\mathrm{T} x^k \geq 0\}$. Set $k := k+1$ and **Goto Step 1**

$\square$

The following two algorithms belong to the class of central cutting plane methods:

- The Elzinga-Moore algorithm [31]: In this method the center is the center of the largest inscribed hypersphere in $\mathcal{P}_k$. A new objective cut simply replaces the old one as the ball-center is not affected by redundant constraints.

- The algorithm of Goffin and Vial [50]: The center is the analytic center (see Section 1.4) of $\mathcal{P}_k$. As the analytic center moves due to redundant constraints, a collection of optimality cuts is kept during the iterations.

A unified treatment of both methods can be found in den Hertog, Kaliski, Roos and Terlaky et al. [55] [2].

In the following we will only consider the Elzinga-Moore method, more precisely we will discuss our variant of the method. The main difference to the Elzinga–Moore method is that instead of solely working with Kelley cuts we employ also the generally deeper Veinott cuts.

For the sake of simplicity of presentation we omit cut-dropping; the same cut-dropping strategies can be used as in the original method. An excellent feature of the Elzinga-Moore method is that it provides lower *and* upper bounds for the optimal objective value which can be utilized in the stopping rule. This also carries over without modifications into our variant; for the sake of a brief presentation we do not include this feature into the algorithm-description below.

The norms appearing below are Euclidean norms, we assume that the objective has been scaled such that $\| c \| = 1$ holds. We will work with a sequence of polyhedrons in $\mathbb{R}^{n+1}$ which will be denoted by $\hat{\mathcal{P}}_1, \hat{\mathcal{P}}_2, \ldots$

Algorithm 1.30 *(Modified Elzinga-Moore central cutting plane method)*

Step 0. *Initialize.* Let $k := 1$, $\underline{f}$ be a lower bound of $c^{\mathrm{T}} x$ on S; $x^0 \in X$; $\hat{\mathcal{P}}_0 := \{(x, \eta) \mid x \in X, \ c^{\mathrm{T}} x - \eta \geq \underline{f}\}$ and let x° be a Slater-point. Let furthermore $\epsilon > 0$ be a stopping tolerance and flag:=true (indicating that Veinott-cuts will be generated).

Step 1. *Compute the "center" of $\mathcal{P}_{k-1}$.* Solve the following linear programming problem: $\max\{\eta \mid (x, \eta) \in \hat{\mathcal{P}}_{k-1}\}$, let a solution be (x^k, η_k).

Step 2. *Test for feasibility and optimality.* If $\eta^k < \epsilon \Longrightarrow$ **STOP**, x_k is our approximate optimal solution. Otherwise: If $x^k \in S$ then **Goto Step 4**, otherwise **Goto Step 3**

Step 3. *Add a feasibility cut.*

- If flag = true then add a Veinott cut (see Section 1.2.2): Let z^k be the intersection of the line segment $[x^\circ, x^k]$ with the boundary of S and i_0 be the index of an active constraint at this point. Update: $\hat{\mathcal{P}}_k = \hat{\mathcal{P}}_{k-1} \cap \{(x, \eta) \mid \nabla^{\mathrm{T}} g_{i_0}(z^k)(x - z^k) - \| \nabla g_{i_0}(z^k) \| \eta \geq 0\}$.

- If flag = false then add a Kelley cut (see Section 1.2.2): Let $i_0 : g_{i_0}(x^k) = \min_i g_i(x^k)$ and $\hat{\mathcal{P}}_k := \hat{\mathcal{P}}_{k-1} \cap \{x \mid g_{i_0}(x^k) + \nabla g_{i_0}(x^k)(x - x^k) - \| \nabla g_{i_0}(x^k) \| \eta \geq 0\}$.

---

[2]Recently the same authors developed an algorithm based on a combination of path–following and cutting planes, [56].

In both cases set $k := k + 1$ and **Goto Step 1**

Step 4. *Add an objective cut.* $\hat{\mathcal{P}}_k$ is obtained from $\hat{\mathcal{P}}_{k-1}$ by replacing the objective cut by $c^T x - \eta \geq c^T x^k$. **Goto Next Step**

Step 5. *Change the Slater-point.* If x^k is a Slater point then set $x^\circ := x^k$ and flag:=true. Otherwise if flag=true and $c^T x^\circ \geq c^T x^k$ holds (i.e. the objective cut does not cut off the Slater-point) then leave x° and flag unchanged. In all other cases set flag:=false. Set $k := k + 1$ and **Goto Step 1**.

$\square$

The ball-centers in the Elzinga-Moore method are computed by utilizing the following result of Nemhauser and Widhelm [119]:

Proposition 1.27 *Let* $\mathcal{P} = \{x \mid a_i^T x \geq b_i, \ i = 1, \ldots, m\}$ *and* $\hat{\mathcal{P}} = \{(x, \eta) \mid a_i^T x - \| a_i \| \eta \geq b_i, \ i = 1, \ldots, m\}$ *Then the center* $\hat{x}$ *and radius* $\hat{\eta}$ *of the largest inscribed hypersphere in* $\mathcal{P}$ *can be computed as:* $(\hat{x}, \hat{\eta}) = \operatorname{argmax}\{\eta \mid (x, \eta) \in \hat{\mathcal{P}}\}$.

Proof: See Nemhauser and Widhelm [119]. $\square$

Convergence behavior

As in [31] let us define a sequence of candidate solutions as follows: Let $z^0 = x^0$ and for $k > 0$ let

$$z^k = \begin{cases} x^k & \text{if } x^k \in S \\ z^{k-1} & \text{otherwise.} \end{cases}$$

Proposition 1.28 *The modified Elzinga-Moore algorithm terminates in a finite number of iterations with an approximate optimal solution. In other words: If we remove the stopping criterion then for the generated sequence the following assertions hold:*

- $\lim_{n \to \infty} \eta_n = 0$.

- $\exists \hat{k}$ *such that for* $k \geq \hat{k}$ $z^k \in S$.

- *Each accumulation point of the sequence* $\{z^k\}_{k=1}^\infty$ *is a solution of Problem (1.76).*

Proof: The proof for the original method as given in Elzinga-Moore [31] goes through also for the modified algorithm. $\square$

1.6 Reduced gradient methods

This section is devoted to algorithms based on the reduced gradient idea of Wolfe [162].

We begin with the case of linear constraints. Let us consider the following problem:

$$
\left.
\begin{aligned}
\max \quad & f(x) \\
A x \ &= b \\
x \ &\geq 0
\end{aligned}
\right\}
\tag{1.77}
$$

where $f : \mathbb{R}^n \mapsto \mathbb{R}^1$, $b \in \mathbb{R}^m$, A is an $m \times n$ matrix and f is concave and continuously differentiable. For the sake of simplicity we assume that A has full row rank. We assume furthermore that the set of optimal solutions is nonempty and bounded. Let S denote the set of feasible solutions.

Let us consider a fixed feasible point x and assume that a partition

$$
x^{\mathrm{T}} = (y, z)^{\mathrm{T}}, \quad A = (B, N)
\tag{1.78}
$$

exists such that $y \in \mathbb{R}^m$, $z \in \mathbb{R}^{n-m}$, B is an $m \times m$ nonsingular matrix and $y > 0$ holds. A partition with these properties will be called a *nondegenerate partition* corresponding to x.

Let $\epsilon > 0$ be given and consider the following direction-finding problem for w partitioned according to the above partitioning as $w^{\mathrm{T}} = (u, v)^{\mathrm{T}}$:

$$
\left.
\begin{aligned}
\max \quad & \tau, \\
\nabla_y^{\mathrm{T}} f(x) u + \nabla_z^{\mathrm{T}} f(x) v \ -\tau &\geq 0, \\
B u + \quad N v \ &= 0, \\
v_j \ &\geq 0, \ \text{if } z_j \leq \epsilon, \\
& \qquad j = 1, \ldots, n - m \\
\| v \| \ &\leq 1.
\end{aligned}
\right\}
\tag{1.79}
$$

Let

$$
J_\epsilon(x) = \{ j \mid 1 \leq j \leq n - m, \ z_j \leq \epsilon \}.
$$

Due to the nonsingularity of B, Problem (1.79) has an optimal solution and the optimal objective value τ^* is clearly nonnegative. If $\tau^* > 0$ then w is a feasible direction along which the objective can locally strictly be increased. If $\tau^* = 0$ then we have no conclusion. If however in this situation we solve (1.79) again but now with $\epsilon = 0$, and $\tau^* = 0$ still holds, then x is an optimal solution of Problem (1.77). The optimality of x follows easily from the Kuhn-Tucker theorem 1.3. The tolerance ϵ plays the role in the algorithm of preventing it from zig-zagging, and it will be reduced during the iteration process.

Wolfe's idea [162] in using a direction-finding subproblem of the above type is the following: Problem (1.79) has an explicit solution with many norms. In fact, u can be eliminated which results in the following equivalent problem:

$$\begin{aligned} \max \quad & r^T v \\ & v_j \quad \geq 0, \text{ if } j \in J_\epsilon(x) \\ & \| v \| \leq 1 \end{aligned} \Bigg\} \tag{1.80}$$

where r is the *reduced gradient* defined by:

$$r^T = \nabla_z^T \mathrm{f}\,(x) - \nabla_y^T \mathrm{f}\,(x) \cdot B^{-1} N$$

and

$$u = -B^{-1} N \, v \tag{1.81}$$

holds. Using the Euclidean norm, the explicit solution of (1.80) can be obtained as follows. Let

$$\tilde{v}_i = \begin{cases} 0 & \text{if } j \in J_\epsilon(x) \text{ and } \dfrac{\partial \mathrm{f}\,(x)}{\partial z_i} < 0, \\[2mm] \dfrac{\partial \mathrm{f}\,(x)}{\partial z_i} & \text{otherwise} \end{cases} \tag{1.82}$$

and $v = \dfrac{\tilde{v}}{\| \tilde{v} \|}$. Notice that the optimal objective value in (1.79) is $\tau^* = \| \tilde{v} \|$.

The reduction, i.e. the equivalence of Problems (1.79) and (1.80) is based on a nondegeneracy assumption, namely that $y > 0$ holds for the partition (1.78). If for a given x and partition (1.78) this is not the case, we can proceed as follows: By pivoting according to a greedy strategy we arrive at a basis for which a maximum number of y_i's is strictly positive. Such a partition will be called a *maximal partition* corresponding to x. If now $y > 0$ holds we are done. Otherwise we utilize the following result due to Kleinmichel and Sadowski [91]: If x is not optimal then there exists an $\epsilon > 0$ and a partition (1.78) such that the direction v determined by solving the reduced problem (1.80) along with the corresponding u solves the full-scale direction-finding problem (1.80) and $\tau^* > 0$ holds. Such a partition can be found either by the lexicographic strategy proposed by Kleinmichel and Sadowski in [91] or by the technique of Dembo and Klincewicz [24].

For the sake of simplicity of presentation we will formulate the reduced gradient method under the assumption that no substantial degeneracy occurs, i.e. the above-mentioned greedy strategy always results in a nondegenerate partition.

Algorithm 1.31 (*Reduced gradient method*)

Step 0. *Initialize.* Let $k := 1$, $\epsilon_1 > 0$. Compute a starting feasible point x^1 by the simplex method. By further pivoting according to a greedy strategy determine a nondegenerate partition $A = (B^1, N^1)$.

Step 1. *Compute the nonbasic components of the direction*   With the current partition $A = (B^k, N^k)$ compute the nonbasic components v^k of the direction vector w^k according to (1.82).

Step 2. *Test for optimality, reduce ϵ_k*   Check whether $\tau^* \geq \epsilon_k$ holds. If yes then set $\epsilon_{k+1} := \epsilon_k$ and **Goto Step 3**. Otherwise:

-   If $\tau^* > 0$ then set $\epsilon_{k+1} := 0.5 \cdot \epsilon_k$ and **Goto Step 3**.

-   Otherwise compute a nonbasic direction $\hat{v}$ according to (1.82) with $\epsilon = 0$; let $\tau^{**} = \| \hat{v} \|$. If $\tau^{**} = 0$ then x^k is optimal, $\Longrightarrow$ **STOP**. Otherwise set $\epsilon_{k+1} := 0.5 \cdot \epsilon_k$ and **Goto Step 1**.

Step 3. *Compute the basic components of the direction*   Compute u^k according to (1.81).

Step 4. *Perform linesearch*   Determine λ_{max} as:

$$\lambda_{max} = \max\{ \lambda \mid x^k + \lambda \cdot w^k \in S \}$$

Due to the nondegeneracy of the partition $\lambda_{max} > 0$ holds. Compute the steplength λ_k according to:

$$\lambda_k = \mathrm{argmax}_\lambda \{ f(x^k + \lambda \cdot w^k) \mid \lambda \in [0, \lambda_{max}] \}$$

Because of $\tau^* > 0$ we have $\lambda_k > 0$.

Step 5. *Perform a step*   Set $x^{k+1} := x^k + \lambda_k \cdot w^k$

Step 6. *Perform pivot steps if necessary*   Check whether the partition (B^k, N^k) is nondegenerate for x^{k+1}. If yes then set $B^{k+1} := B^k$, $N^{k+1} := N^k$, $k := k + 1$ and **Goto Step 1**. Otherwise by greedy pivoting determine a maximal partition (B^{k+1}, N^{k+1}) to x^{k+1}. According to our nondegeneracy assumption this will be a nondegenerate partition. Set $k := k + 1$ and **Goto Step 1**.

$\square$

The algorithm as formulated above is a feasible direction method closely related to Zoutendijk's P1 method [165]. In fact, replacing the norm-condition in (1.79) by $\| (u, v) \| \leq 1$ and determining the direction w in the algorithm by solving the modified direction-finding problem (1.79), Zoutendijk's P1 method results. This close connection suggests that the convergence behavior of the two methods is very similar. From this point of view the main difference is that in the reduced gradient method the norm may change from iteration to iteration due to changing partitions. This is no real problem however because the number of different partitions is finite.

Wolfe originally formulated his algorithm in [162] with $\epsilon = 0$. In this form the algorithm does not converge in general, see Wolfe [163]. The idea of using Zoutendijk's ϵ_k-anti-zigzagging technique to obtain a convergent version of the reduced gradient method is due to Kleinmichel and Sadowski [91]. The algorithm above, published in [91], is one of the first convergent versions of the reduced gradient method, and definitely the first reduced gradient type method which also works in the degenerate case. In Step 4 (linesearch) the minimization along the line can be replaced by other stepsize-rules, see e.g. Grossmann and Kleinmichel [53].

Convergence behavior

Proposition 1.29 *If the algorithm does not stop at Step 2 with an optimal solution, then all accumulation points of the sequence of points generated by the algorithm are solutions of Problem (1.77).*

Proof: See Kleinmichel and Sadowski [91].   □

Let us remark that the same proposition holds if degeneracy is allowed, when considering the algorithm endowed with a degeneracy-handling subcyle, see the paper cited.

The main drawback of the method as formulated above is that it is the constrained analogue of the steepest descent method of unconstrained optimization and thus it has a similarly poor convergence rate. The idea of overcoming this difficulty is due to Murtagh and Saunders [117] and is implemented in MINOS. Very shortly it can be summarized as follows. The partition now consists of triplets:

$$A = (B, S, N), \quad x = (x_B, x_S, x_N) \text{ and } w = (w_B, w_S, w_N)$$

with the property $x_B > 0$, $x_S > 0$, $x_N = 0$. The components corresponding to x_B, x_S, x_N are called basic-, superbasic- and nonbasic components, respectively. The nonbasic components are kept on zero level by fixing $w_N = 0$. Under these circumstances the reduced direction-finding subproblem (1.80) is explicitly solvable using ellipsoid norms. This implies that as long as S is not changed powerful quasi-Newton methods can be used in the corresponding subcycle of iterations. If a superbasic variable hits a bound it will be qualified as nonbasic. If a basic variable hits a bound it is exchanged against a superbasic variable as long as this is possible, and the outgoing variable will be qualified as nonbasic. If there are no more superbasic variables or the above outlined change of the basis is not possible, then a steepest ascent step is made by using the original direction-finding subproblem (1.80) of the reduced gradient method, resulting in a new triplet of partition.

Let us next consider the following nonlinearly constrained problem:

$$\left. \begin{aligned} \max \quad & f(x) \\ & G(x) \geq 0 \\ & A x = b \\ & x \geq 0 \end{aligned} \right\} \tag{1.83}$$

where $f : \mathbb{R}^n \mapsto \mathbb{R}^1$, $G : \mathbb{R}^n \mapsto \mathbb{R}^q$, $G(x)^T = (g_1(x), \ldots, g_q(x))$, $b \in \mathbb{R}^m$, A is an $m \times n$ matrix. We assume that f and g_i $\forall i$ are concave, continuously differentiable functions. For the sake of simplicity we assume that A has full row rank. We assume furthermore that the set of optimal solutions is nonempty and bounded. Let S denote the set of feasible solutions.

The first application of the reduced gradient idea to the nonlinearly constrained case is due to Abadie and Carpentier [1]. They developed GRG, a **G**eneralized **R**educed **G**radient method for this type of problems with the nonlinear constraints formulated as equalities. The general idea is the following: Assume that we have a feasible point x. The nonlinear constraints are linearized by employing the truncated Taylor-expansion and a direction is determined according to the reduced gradient method for the linear case. After making a linesearch the new point is in general not feasible. For returning to the feasible surface Newton's method is used.

The rest of this section is devoted to a reduced gradient type algorithm which works with feasible points. We present it with Zoutendijk's ϵ_k anti-zig-zag technique much in the same fashion as for the linear case. Let us assume that for the linear part we have a nondegenerate partition (1.78) at our current feasible point x. The direction-finding subproblem can then be formulated as follows:

$$
\left.
\begin{aligned}
\max \quad & \tau, \\
& \nabla_y^T f(x) u + \nabla_z^T f(x) v \quad - \tau \geq 0, \\
& \nabla_y^T g_k(x) u + \nabla_z^T g_k(x) v - \theta_k \tau \geq 0, \text{ if } g_k(x) \leq \epsilon \\
& \hspace{6cm} k = 1, \ldots, q \\
& B u + \quad N v \quad = 0, \\
& \hspace{2.5cm} v_j \quad \geq 0, \text{ if } z_j \leq \epsilon, \\
& \hspace{4cm} j = 1, \ldots, n - m \\
& \hspace{2cm} \| v \| \quad \leq 1
\end{aligned}
\right\}
\tag{1.84}
$$

where $\theta^T = (\theta_1, \ldots, \theta_q)^T$ is a vector of positive weights. Let ∇G denote the Jacobi-matrix of the nonlinearly constrained part (the constraint gradients are in the rows) and let us introduce the following index-set:

$$K_\epsilon(x) = \{\, j \mid 1 \leq j \leq q, \ g_j(x) \leq \epsilon \,\}.$$

Carrying out the same reduction (1.81) as for the linear case results in the following reduced direction-finding subproblem for the nonbasic part of the direction:

$$
\left.
\begin{aligned}
\max \quad & \tau, \\
& r^T v \quad - \tau \geq 0 \\
& t_k^T(x) v \quad - \theta \tau \geq 0, \text{ if } k \in K_\epsilon(x) \\
& \hspace{1cm} v_j \quad \geq 0, \text{ if } j \in J_\epsilon(x) \\
& \hspace{0.5cm} \| v \| \quad \leq 1
\end{aligned}
\right\}
\tag{1.85}
$$

where r is the *reduced gradient* as defined above and $T(x) = (t_1(x), \dots, t_q(x))^{\mathrm{T}}$ is the reduced Jacobi-matrix defined by:

$$T(x) = \nabla_z \mathrm{G}\,(x) - \nabla_y \mathrm{G}\,(x) \cdot B^{-1}N.$$

Notice that (1.85) is no more explicitly solvable when q > 0. Nevertheless it is a linear programming problem if we choose the $\| \cdot \|_\infty$ norm.

The direction finding problem (1.84) is, except for the norm-condition, identical with the direction finding problem of Zoutendijk's P1 method. Working with partitions of variables and choosing the norm in (1.84) as in the reduced gradient method results in a different direction. For being able to solve efficiently the reduced direction finding problem (1.85), the ∞-norm is choosen there. This introduces a drawback when comparing the method with the reduced gradient method for linear constraints: Most of the components of the nonbasic direction are now 1, 0 or -1, thus reflecting only roughly the direction of local increase in the nearly quadratic case.

Let us assume again that no substantial degeneracy occurs; for the degenerate case the same comments apply as for the linear case. The algorithm can now be formulated exactly in the same manner as for the linear case except that now the nonbasic directions are computed according to (1.85) and a Phase-I procedure is needed which accounts for the nonlinear constraints. We specify the algorithm below in detail for the sake of completeness.

Algorithm 1.32 *(A reduced gradient feasible directions method)*

Step 0. *Initialize.* Let $k := 1$, $\epsilon_1 > 0$. Compute a starting feasible point x^1 and determine a nondegenerate partition $A = (B^1, N^1)$ by a Phase-I strategy.

Step 1. *Compute the nonbasic components of the direction* With the current partition $A = (B^k, N^k)$ compute the nonbasic components v^k of the direction vector w^k by solving (1.85).

Step 2. *Test for optimality, reduce ϵ_k* Check whether $\tau^* \geq \epsilon_k$ holds. If yes then set $\epsilon_{k+1} := \epsilon_k$ and **Goto Step 3**. Otherwise:

- If $\tau^* > 0$ then set $\epsilon_{k+1} := 0.5 \cdot \epsilon_k$ and **Goto Step 3**.

- Otherwise compute a nonbasic direction $\hat{v}$ according to (1.85) with $\epsilon = 0$; let $\tau^{**} = \| \hat{v} \|$. If $\tau^{**} = 0$ then x^k is optimal, $\Longrightarrow$ **STOP**. Otherwise set $\epsilon_{k+1} := 0.5 \cdot \epsilon_k$ and **Goto Step 1**.

Step 3. *Compute the basic components of the direction* Compute u^k according to (1.81).

Step 4. *Perform linesearch* Determine λ_{max} as:

$$\lambda_{max} = \max\{\, \lambda \mid x^k + \lambda \cdot w^k \in S \,\}$$

Due to the nondegeneracy of the partition $\lambda_{max} > 0$ holds. Compute the steplength λ_k according to:

$$\lambda_k = \mathrm{argmax}_\lambda \{\, \mathrm{f}\,(x^k + \lambda \cdot w^k) \mid \lambda \in [0, \lambda_{max}] \,\}$$

Because of $\tau^* > 0$ we have $\lambda_k > 0$.

Step 5. *Perform a step*   Set $x^{k+1} := x^k + \lambda_k \cdot w^k$

Step 6. *Perform pivot steps if necessary*   Check whether the partition (B^k, N^k) is nondegenerate according to x^{k+1}. If yes then set $B^{k+1} := B^k$, $N^{k+1} :=$ N^k, $k := k+1$ and **Goto Step 1**. Otherwise by greedy pivoting determine a maximal partition (B^{k+1}, N^{k+1}) to x^{k+1}. According to our nondegeneracy assumption this will be a nondegenerate partition. Set $k := k+1$ and **Goto Step 1**.

□

For convex programming problems the algorithm was first proposed and theoretically studied by Sadowski [143]; for chance constrained problems see also the note in Section 3.2.

Convergence behavior

The algorithm behaves from the convergence point of view similarly as the reduced gradient method for linear constraints; for a convergence theorem see Sadowski [143].

A Phase-I procedure can be designed according to general Phase-I schemes of nonlinear programming e.g. as follows: Let us assume that our problem is Slater-regular. Set up an auxiliary problem where the sum of the nonlinear constraint functions is maximized subject to the linear constraints. Start up the algorithm. Due to the assumption, after a finite number of iterations at least one of the constraints will be fulfilled. Modify the problem at such iterations by removing the feasible constraint functions from the objective and by appending them to the constraints.

Chapter 2

Stochastic linear programming models

This chapter serves for presenting the stochastic linear programming (SLP) model classes considered in this work. We do not intend to present a classification of SLP models, we only discuss those SLP models for which solvers are available for us and thus they are included into the computational study. The first section deals with two-stage models followed by a section devoted to the discussion of jointly chance-constrained models. In the final section inequalities playing an important role in SLP-algorithms are presented, along with their computational algorithms.

2.1 Two stage models

In this section the main properties of two-stage models will be summarized, for detailed presentations see Kall [71], Kall and Wallace [85] and Wets [161].

Two-stage fixed recourse models can be formulated as follows:

$$
\left.\begin{array}{rl}
\min \left[c^{\mathrm{T}}x + E_\omega Q(x, \xi(\omega)) \right] & \\
A\,x & = b \\
x & \geq 0
\end{array}\right\} \tag{2.1}
$$

where

$$
\left.\begin{array}{rl}
Q(x, \xi(\omega)) = \min q^{\mathrm{T}}(\xi(\omega))y & \\
W\,y & = h(\xi(\omega)) - T(\xi(\omega))x \\
y & \geq 0.
\end{array}\right\} \tag{2.2}
$$

and A is an $(m_1 \times n_1)$ whereas W an $(m_2 \times n_2)$ matrix, respectively. The dimensions of all of the other arrays are fixed accordingly.

Matrix W is called the recourse matrix. In the special case when the recourse matrix has the form W=(I,-I), (2.1) is called a simple recourse problem.

In the models above $\omega \in \Omega$, $(\Omega, \mathcal{F}, P)$ is a probability space, and the random entries in the various arrays are represented through affine sums

$$\left.\begin{array}{rcl} q(\xi(\omega)) & = & \bar{q}^0 + \sum_{j=1}^k \bar{q}^j \xi_j(\omega) \\ h(\xi(\omega)) & = & \bar{h}^0 + \sum_{j=1}^k \bar{h}^j \xi_j(\omega) \\ T(\xi(\omega)) & = & \bar{T}^0 + \sum_{j=1}^k \bar{T}^j \xi_j(\omega) \end{array}\right\} \tag{2.3}$$

with $\xi_j(\omega)$, j=1,..k being random variables. The support of the random vector $\xi(\omega)^{\mathrm{T}} = (\xi_1(\omega), \ldots, \xi_k(\omega))$ will be denoted by Ξ, i.e. Ξ is the smallest closed set in $\mathbf{R}^k$ for which $P(\{\omega \mid \xi(\omega) \in \Xi\}) = 1$ holds.

The LP-dual of the second-stage problem reads as

$$\left.\begin{array}{c} \max (h(\xi(\omega)) - T(\xi(\omega))x)^{\mathrm{T}} u \\ W^{\mathrm{T}} u \ \leq q(\xi(\omega)) \end{array}\right\} \tag{2.4}$$

Let us introduce the following denotations:

$$\begin{array}{l} \mathcal{Q}(x) = E_\omega Q(x, \xi(\omega)) = \int_\Omega Q(x, \xi(\omega)) \, P(d\omega) \\ \psi(x) = c^{\mathrm{T}} x + \mathcal{Q}(x) \\ X \quad = \{x \mid Ax = b, x \geq 0\} \\ K \quad = \{x \mid (2.2) \text{ is feasible w.p. } 1\}. \end{array} \tag{2.5}$$

It is well-known that K is a closed convex set, and under suitable integrability conditions on $\xi(\omega)$ the integral involved in the definition of $\mathcal{Q}(x)$ exists on K, see Kall [71]. Problem (2.1) can be reformulated as follows:

$$\left.\begin{array}{c} \min [c^{\mathrm{T}}x + \mathcal{Q}(x) \\ x \in X \cap K \end{array}\right\} \tag{2.6}$$

In this work we will consider *complete recourse problems*. Let us mention that in the case when $\xi(\omega)$ has a finite discrete distribution the integrability problem does not arise and some of the algorithms are able to handle the *induced feasibility set* K. We will make the following assumptions:

- $\xi(\omega) \in \mathcal{L}^2(P)$.

- W is a complete recourse matrix, i.e.

$$\{y \mid Wy = z, y \geq 0\} \neq \emptyset, \ \forall z \in \mathbf{R}^{m_2}.$$

- An optimal solution of the second stage problem exists almost surely,

$$\{u \mid W^{\mathrm{T}}u \leq q(\xi(\omega))\} \neq \emptyset \text{ w.p. } 1.$$

Proposition 2.1 *Under the above assumptions the following assertions hold:*

- $K = {\rm I\!R}^{n_1}$ *and* $\mathcal{Q}(x)$ *is finite* $\forall x \in {\rm I\!R}^{n_1}$.

- *If* $q(\xi(\omega)) \equiv q^0$ *then for any fixed* $x \in X$:

$$\xi \mapsto Q(x, \xi) \text{ is piecewise linear and convex}$$

- *If* $h(\xi(\omega)) \equiv h^0$ *and* $T(\xi(\omega)) \equiv T^0$ *then for any fixed* $x \in X$:

$$\xi \mapsto Q(x, \xi) \text{ is piecewise linear and concave}$$

- *For any fixed* ξ:

$$x \mapsto Q(x, \xi) \text{ is piecewise linear and convex on } X$$

-

$$\mathcal{Q}(x) \text{ is finite and convex on } X$$

Proof: See e.g. Kall [71] or Wets [161].   □

Let us formulate for the case $q(\xi(\omega)) \equiv q$ the *expected value problem*. Substituting $\xi(\omega)$ in the second and third relation of (2.3) by $E(\xi(\omega))$, we denote the resulting arrays by $\hat{h}, \hat{T}$, respectively. The expected value problem is the following LP problem:

$$\left.\begin{array}{rl} \min\, [\, c^\mathrm{T}x + q^\mathrm{T}y\,] & \\ A\,x \quad\quad\quad = b & \\ \hat{T}x + \quad W\,y = \hat{h} & \\ x \quad\quad\quad\quad \geq 0 & \\ y \quad \geq 0. & \end{array}\right\} \tag{2.7}$$

Let us consider the general case (i.e. $q(\xi)$) when $\xi(\omega)$ has a finite discrete distribution specified by:

$$\begin{pmatrix} p_1 & p_2 & \cdots & p_L \\ \xi^1, & \xi^2 & \cdots & \xi^L \end{pmatrix}$$

where

$$p_l = P(\{\omega \mid \xi(\omega) = \xi^l\}), \; l = 1, \ldots, L$$

$$p_l > 0, \; l = 1, \ldots, L; \; \sum_{l=1}^{L} p_l = 1.$$

Denoting by (q^l, h^l, T^l) the realizations resulting from the substitution of ξ^l into the affine relations (2.3) we get the following distribution for the random entries:

$$\begin{pmatrix} p_1 & p_2 & \cdots & p_L \\ (q^1, h^1, T^1) & (q^2, h^2, T^2) & \cdots & (q^L, h^L, T^L) \end{pmatrix}$$

In the case of finite discrete distributions (2.1) can equivalently be reformulated as the following algebraic equivalent LP (algebraic means that the problem can entirely be formulated in algebraic terms):

$$
\left.\begin{array}{rl}
\min \left[c^T x + p_1 (q^1)^T y^1 + \ldots + p_L (q^L)^T y^L \right] & \\
A\,x \qquad\qquad\qquad\qquad\qquad\qquad\qquad\quad = b & \\
T^1 x + \quad W\,y^1 \qquad\qquad\qquad\qquad\; = h^1 & \\
\vdots \qquad\qquad\qquad \ddots \qquad\qquad\qquad \vdots & \\
T^L x + \qquad\qquad\qquad\qquad W\,y^L \quad = h^L & \\
x \qquad\qquad\qquad\qquad\qquad\qquad \geq 0 & \\
y^i \qquad\quad \geq 0\; \forall i. &
\end{array}\right\}
\qquad (2.8)
$$

The structure of this LP can be seen on Figure 2.1

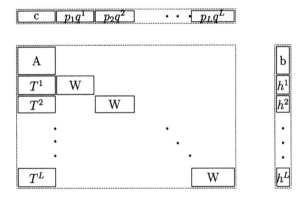

Figure 2.1: The structure of the equivalent LP problem

For a fixed x Problem (2.8) can obviously be decomposed into L subproblems having the form

$$
\left.\begin{array}{rl}
Q(x, \xi^i) = \min\; (q^i)^T y & \\
W\,y = h^i - T^i x & \\
y \geq 0. &
\end{array}\right\}
\qquad (2.9)
$$

The dual of Problem (2.8) is

$$\max \left[\mathbf{b}^T u + (\mathbf{h}^1)^T v^1 + \ldots + (\mathbf{h}^L)^T v^L \right]$$

$$\left.\begin{array}{rl} \mathbf{A}^T u + (\mathbf{T}^1)^T v^1 + \ldots + (\mathbf{T}^L)^T v^L & \le \mathbf{c} \\ \mathbf{W}^T v^1 & \le p_1 \mathbf{q}^1 \\ & \vdots \\ \mathbf{W}^T v^L & \le p_L \mathbf{q}^L. \end{array}\right\} \qquad (2.10)$$

Let us notice that in the finite discrete case the last assertion in Proposition 2.1 can be sharpened:

Proposition 2.2 $\mathcal{Q}(x)$ *is piecewise linear and convex.*

Proof: The assertion follows immediately from the following explicit formula:

$$\mathcal{Q}(x) = \sum_{l=1}^{L} p_l Q(x, \xi^l)$$

$\square$

Let us now consider the case when $T(\xi(\omega)) \equiv T$ and $q(\xi(\omega)) \equiv q$ hold. In this case Problem 2.1 can be reformulated as follows (We use the same symbols for the functions in the objective for the sake of convenience):

$$\left.\begin{array}{rl} \min \left[c^T x + E_\omega Q(\chi, h(\omega)) \right] \\ A x & = b \\ T x - \chi & = 0 \\ x & \ge 0 \end{array}\right\} \qquad (2.11)$$

where

$$\left.\begin{array}{rl} Q(\chi, h(\omega)) = \min & q^T(\xi(\omega))y \\ W y & = h(\omega) - \chi \\ y & \ge 0 \end{array}\right\} \qquad (2.12)$$

where we introduced the abbreviation $h(\omega) = h(\xi(\omega))$. Let

$$\begin{array}{rl} \mathcal{Q}(\chi) & = E_\omega Q(\chi, h(\omega)) = \int_\Omega Q(\chi, h(\omega)) \, P(d\omega) \\ \psi(x, \chi) & = c^T x + \mathcal{Q}(\chi) \\ X & = \{(x, \chi) \mid Ax = b, Tx - \chi = 0, x \ge 0\}. \end{array} \qquad (2.13)$$

The rest of this section is devoted to the case when $T(\xi(\omega)) \equiv T$ and $q(\xi(\omega)) \equiv q$ hold and Problem 2.11 is a simple-recourse problem. With the denotations

$$y = (y^+, y^-) \text{ and } q = (q^+, q^-)$$

the second-stage problem (2.12) gets the following form:

$$Q(\chi, h(\omega)) = \min (q^+)^T y^+ + (q^-)^T y^-$$
$$\left.\begin{array}{rcl} y^+ & - & y^- = h(\omega) - \chi \\ y^+, & & y^- \geq 0. \end{array}\right\} \qquad (2.14)$$

Proposition 2.3 *The second-stage problem (2.14) has an optimal solution $\Longleftrightarrow$ $q^+ + q^- \geq 0$ holds.*

Proof: Follows readily from LP duality.   □

Let us assume in the sequel that the condition in the previous proposition holds. Next we observe that due to separability in (2.14), we have:

$$Q(\chi, h(\omega)) = \sum_{j=1}^{m2} Q_j(\chi_j, h_j(\omega))$$

with

$$Q_j(\chi_j, h_j(\omega)) = \min \quad q_j^+ y_j^+ + q_j^- y_j^-$$
$$\left.\begin{array}{rcl} y_j^+ & - & y_j^- = h_j(\omega) - \chi_j \\ y_j^+, & & y_j^- \geq 0. \end{array}\right\} \qquad (2.15)$$

Observe that (2.15) has an explicit solution which results in the formula:

$$Q_j(\chi_j, h_j(\omega)) = \begin{cases} q_j^+(h_j(\omega) - \chi_j) & \text{if } h_j(\omega) \geq \chi_j \\ q_j^-(\chi_j - h_j(\omega)) & \text{otherwise.} \end{cases} \qquad (2.16)$$

By computing the expected values for the terms we arrive at the following separable form of the second-stage objective (see e.g. Kall [71] and Kall, Ruszczyński and Frauendorfer [83]).

$$\mathcal{Q}(\chi) = \sum_{j=1}^{m2} \mathcal{Q}_j(\chi_j) \qquad (2.17)$$

where

$$\mathcal{Q}_j(\chi_j) = q_j^+(h_j^+(\chi_j) - \chi_j)p_j^+(\chi_j) + q_j^-(\chi_j - h_j^-(\chi_j))p_j^-(\chi_j) \qquad (2.18)$$

with

$$\left.\begin{array}{l} h_j^+(\chi_j) = E(h_j(\omega) \mid h_j(\omega) \geq \chi_j) \\ p_j^+(\chi_j) = P(h_j(\omega) \geq \chi_j) \\ h_j^-(\chi_j) = E(h_j(\omega) \mid h_j(\omega) < \chi_j) \\ p_j^-(\chi_j) = P(h_j(\omega) < \chi_j). \end{array}\right\} \qquad (2.19)$$

Notice that the simple recourse problem with solely the RHS being stochastic only depends on the one-dimensional marginal distributions of $h_j(\xi(\omega))$. This implies that the terms in (2.17) can be handled separately, the j'th term only depending on the (marginal) distribution of $h_j(\xi(\omega))$.

If in the simple-recourse problem we introduce the requirement that the components of y should be integers, we get a problem with *simple integer recourse*, see Klein Haneveld, Stougie and van der Vlerk [90], Louveaux and van der Vlerk [97] and van der Vlerk [157].

2.2 Chance constrained models

This section is devoted to give a brief overview on jointly chance constrained models. For more detailed presentations see Kall [71], Kall and Wallace [85], Mayer [109], Prékopa [123], [125], [130].

Chance constrained (or probabilistic constrained) models can be formulated as follows:

$$\left.\begin{array}{rl} \min & c^T x \\ P(\{\omega \mid T^k x \geq h^k(\omega)\}) & \geq \alpha_k, \ k = 1, \ldots, S \\ A x & = b \\ x & \geq 0 \end{array}\right\} \tag{2.20}$$

where A is an $(m_1 \times n_1)$ whereas T^k an $(m_2^k \times n_1)$, $k = 1, \ldots, S$ matrix, respectively. The dimensions of all of the other arrays are fixed accordingly. The α_k, with $0 \leq \alpha_k \leq 1 \ \forall k$, are given probability levels and

$$h^k(\omega) = \bar{h}^{k,0} + \sum_{j=1}^{r} \bar{h}^{k,j} \xi_j(\omega) \tag{2.21}$$

where $\xi_j(\omega)$, $j = 1, \ldots, r$ are random variables on the probability space $(\Omega, \mathcal{F}, P)$.

If $m_2^k = 1 \ \forall k$ holds then (2.20) is called a stochastic programming model with separate chance constraints, otherwise
it is a jointly chance constrained problem.

Notice that in the model formulation above we have deterministic technology matrices. The reason is that in the jointly chance constrained case this is needed to ensure convexity of (2.20) for a broad class of probability distributions. In the separately chance constrained case, for certain classes of distributions and appropriate probability levels, random entries are allowed also in the matrix meaning that convex programming problems result. This class of distributions includes e.g. the multinormal distribution, see e.g. Kall [71] and Marti [102].

In the algorithms we will make use of the following obvious reformulation

$$\left.\begin{aligned}
\min c^{\mathrm{T}}x \\
P(\{\omega \mid y^k \geq h^k(\omega)\}) \geq \alpha_k, \; k = 1, \ldots, S \\
T^k x - \quad y^k \qquad\qquad \geq 0, \; k = 1, \ldots, S \\
A\,x \qquad\qquad\qquad = b \\
x \qquad\qquad\qquad \geq 0.
\end{aligned}\right\} \qquad (2.22)$$

Introducing the notation $F_k(y^k)$ for the joint probability distribution function of the random vector $h^k(\omega)$, $\forall k$ leads to the equivalent form

$$\left.\begin{aligned}
\min c^{\mathrm{T}}x \\
F_k(y^k) \geq \alpha_k, \; k = 1, \ldots, S \\
T^k x - \quad y^k \geq 0, \; k = 1, \ldots, S \\
A\,x \qquad = b \\
x \qquad \geq 0.
\end{aligned}\right\} \qquad (2.23)$$

The key to the convexity properties of (2.23) is the following theorem:

Proposition 2.4 *(Prékopa [121])*
Assume that the probability distributions of $h_k(\omega)$ $\forall k$ are absolutely continuous. $h_k(\omega)$ has a logarithmic concave (logconcave) density function $\Longleftrightarrow$ the distribution function $F_k(y^k)$ is logconcave.

Proof: Prékopa [121]. □

The above proposition has been generalized to certain classes of quasiconcave measures by Borell [14], Brascamp and Lieb [16] and Rinott [134].

Notice that having quasiconcave distributions, Problem (2.23) is a convex programming problem in the sense that the feasible set is convex. In this study we will only consider the logconcave case. In the numerical computations the distribution will always be non-degenerate multinormal, and hence obviously logconcave.

2.3   Bounds

The purpose of this section is to summarize bounds for both model classes discussed in the previous Chapter. We also discuss the computational algorithms for these bounds. For overviews see Birge and Wets [12], Kall [76], Kall, Ruszczyński and Frauendorfer [83], Kall and Wallace [85], Prékopa [130] and for details the references therein. In this section

$q(\xi(\omega)) \equiv q$ will be assumed throughout.

2.3.1 Lower bounds for two-stage models

The simplest lower bound follows directly from duality in linear programming; it provides a lower bound on the second-stage objective function $Q(x, \xi)$.

Proposition 2.5

- *For any fixed feasible solution u of the second stage dual problem (2.4) the following inequality holds:*

$$(h(\xi) - T(\xi)x)^{\mathrm{T}}u \le Q(x, \xi), \ \forall x \in X, \ \forall \xi \in \Xi$$

- *Let $\xi^* \in \Xi$ and $x^* \in X$ be fixed and let u be an optimal solution of the corresponding second-stage dual (2.4). In this case the inequality above holds as equality in the point (x^*, ξ^*), i.e. the bilinear function on the left hand side reduces to a supporting hyperplane for any fixed x in ξ and vice versa, for the epigraph of $Q(x, \xi)$.*

Proof: Both assertions trivially follow from LP-duality and from the fact that the feasible domain of the second stage dual neither depends on x nor on ξ. □

The inequalities in the rest of this subsection utilize the convexity of the function $Q(x, \xi)$ in ξ, for any fixed $x \in X$. We work with a closed convex set containing the support. For notational simplicity we will denote this set again with Ξ. The convexity directly implies the following inequality:

Proposition 2.6 *(Jensen inequality) Let $x \in X$ be fixed and $\xi^i \in \Xi, \lambda_i \ge 0, i = 1, \ldots, N,$ $\sum_{i=1}^{N} \lambda_i = 1$. Then:*

$$Q(x, \sum_{i=1}^{N} \lambda_i \xi_i) \le \sum_{i=1}^{N} \lambda_i Q(x, \xi_i)$$

Proof: Trivial. □

Let us now turn our attention to the integral form of the Jensen-inequality. For later usage in algorithms we formulate the bound for a partition of Ξ. Let us introduce the following denotations:

$$\Xi \quad = \bigcup_{l=1}^{L} \Xi_l; \; \Xi_l \cap \Xi_k = \emptyset \; (l \neq k)$$

Ξ_l is a convex Borel-set , $l = 1, \ldots, L$.

$\Omega_l \quad = $ inverse image of Ξ_l under ξ.

$p_l \quad = P(\{\omega \mid \xi(\omega) \in \Xi_l\})$ $\hspace{3cm}$ (2.24)

$\mathcal{C} \quad = \sigma(\Xi_1, \ldots, \Xi_L)$ the σ algebra generated by $\Xi_1, \ldots, \Xi_L$

$\tilde{\xi}(\omega) \quad = E^{\mathcal{C}}(\xi(\omega))$ the conditional expectation w.r. to $\mathcal{C}$

$\tilde{\xi}^l \quad = E_{\Xi_l}(\xi) = \frac{1}{P(\Xi_l)} \int_{\Xi_l} \xi P(d\xi), \; \forall l$

Proposition 2.7 *(Jensen inequality, integral form) For any $x \in X$ the following inequality holds:*

$$Q(x, E^{\mathcal{C}}(\xi)) \leq E^{\mathcal{C}}(Q(x, \xi)).$$

Proof: By Proposition 2.1 the function $Q(x, \xi)$ is convex in ξ. Our assertion follows from Jensen's inequality for conditional expectations, see Pfanzagl [120]. $\hspace{1cm}$ □

Notice that $\tilde{\xi}(\omega)$ takes on the constant values $\tilde{\xi}^l$ on Ω_l $\forall l$ thus having a finite discrete distribution.

Proposition 2.8 *For any $x \in X$ the following inequalities hold:*

$$\tilde{Q}_l(x) \stackrel{\text{def}}{=} p_l Q(x, \tilde{\xi}^l) \leq \int_{\Xi_l} Q(x, \xi) P(d\xi), \; \forall l. \hspace{2cm} (2.25)$$

$$\tilde{Q}(x) \stackrel{\text{def}}{=} \sum_{l=1}^{L} \tilde{Q}_l(x) \leq Q(x). \hspace{2cm} (2.26)$$

Proof: The assertion follows directly from the previous Proposition. $\hspace{1cm}$ □

Proposition 2.9 *Assume that for a fixed x, $Q(x, \xi)$ is an affine function on Ξ_l. Then the approximate and exact values coincide, i.e.*

$$p_l Q(x, \tilde{\xi}^l) = \int_{\Xi_l} Q(x, \xi) P(d\xi) \hspace{2cm} (2.27)$$

holds.

Proof: Follows directly by computing the integral in the relation (2.25), for an affine function. $\hspace{1cm}$ □

Let us consider the following problem:

$$\left. \begin{array}{lr} \min c^{\mathrm{T}} x + \sum_{l=1}^{L} p_l Q(x, \tilde{\xi}^l) & \\ A\,x \hspace{3cm} = b \\ x \hspace{3.5cm} \geq 0, \end{array} \right\} \hspace{1.5cm} (2.28)$$

As a consequence of Proposition 2.8, the optimal objective value of (2.28) is a lower bound for the optimal objective value of our two-stage problem (2.2). On the other hand, (2.28) is a two-stage problem itself with a finite discrete distribution. We can interpret this situation as follows: The random variable $\xi(\omega)$ has been approximated by the discretely distributed random variable $\tilde{\xi}(\omega)$ such that the optimal objective value yields a lower bound to the true value. Let us notice that for L=1 we get that the optimal objective value of the expected-value problem (2.7) is a lower bound for the true value.

Let us now consider the inequalities above for refined partitions. Let $\Xi'_1, \ldots, \Xi'_K$ be a second partition with the properties specified above, which is a refinement of $\Xi_1, \ldots, \Xi_L$. Denotations:

$$
\begin{aligned}
p'_l &= P(\{\omega \mid \xi(\omega)) \in \Xi'_l\}) \\
\mathcal{D} &= \sigma(\Xi'_1, \ldots, \Xi_K)' \\
\check{\xi}(\omega) &= E^{\mathcal{D}}(\xi(\omega)) \text{ the conditional expectation w.r. to } \mathcal{D}.
\end{aligned}
\tag{2.29}
$$

Obviously $\mathcal{C} \subset \mathcal{D}$ holds. Let us denote the realizations of the finitely distributed $\check{\xi}$ by $\check{\xi}^1, \ldots, \check{\xi}^K$.

Proposition 2.10 *The lower bound resulting from the finer subdivision is at least as good as the previous one, i.e.*

$$
\sum_{l=1}^{L} p_l Q(x, \tilde{\xi}^l) \leq \sum_{k=1}^{K} p'_k Q(x, \check{\xi}^k)
$$

Proof: Just to show how this type of proofs run, let us carry it out in detail. Assume that

$$
\Xi_l = \bigcup_{k \in I_l} \Xi'_k
\tag{2.30}
$$

holds. We shall prove that

$$
p_l Q(x, \tilde{\xi}^l) \leq \sum_{k \in I_l} p'_k Q(x, \check{\xi}^k)
$$

holds. This is equivalent with:

$$
Q(x, \tilde{\xi}^l) \leq \sum_{k \in I_l} \frac{p'_k}{p_l} Q(x, \check{\xi}^k)
$$

By observing $\sum_{k \in I_l} p'_k = p_l$ and utilizing the convexity of $Q(x, \xi)$ we get:

$$
\begin{aligned}
\sum_{k \in I_l} \frac{p'_k}{p_l} Q(x, \check{\xi}^k) &\geq Q(x, \frac{1}{p_l} \sum_{k \in I_l} p'_k \check{\xi}^k) \\
&= Q(x, \frac{1}{p_l} \sum_{k \in I_l} p'_k \frac{\int_{\Xi'_k} \xi P(d\xi)}{p'_k}) \\
&= Q(x, \frac{1}{p_l} \int_{\Xi_l} \xi P(d\xi)) \\
&= Q(x, \tilde{\xi}^l).
\end{aligned}
$$

Computation

Notice that (2.28) is a two-stage problem with a finite discrete distribution with the number of realizations equal to the number of subsets of the subdivision. It can be solved by any algorithm devised for these type of problems. For later use let us formulate the dual of the LP-equivalent of Problem (2.28):

$$\left.\begin{array}{c} \max b^{\mathrm{T}}u + \sum_{l=1}^{L} h(\tilde{\xi}^l)^{\mathrm{T}}v^l \\ A^{\mathrm{T}}u + \sum_{l=1}^{L} T(\tilde{\xi}^l)^{\mathrm{T}}v^l \le c, \\ W^{\mathrm{T}}v^l \le p_l q, \ l = 1, \ldots, L. \end{array}\right\} \tag{2.31}$$

2.3.2 Upper bounds for two-stage models

Let Ξ be an interval containing the support, i.e.

$$\Xi = \prod_{i=1}^{r}[a_i, b_i]$$

We consider first the case r=1, i.e. $\Xi = [a, b]$ and a partition $\Xi_1, \ldots, \Xi_L$ of Ξ consisting now of subintervals

$$[t^1, t^2), \ldots, [t^L, t^{L+1}], \text{ with } t^1 = a \text{ and } t^{L+1} = b.$$

The following inequalities hold.

Proposition 2.11 *(Edmundson-Madansky inequality, [100])*

$$\int_{\Xi_l} Q(x, \xi)P(d\xi) = \int_{t_l}^{t_{l+1}} Q(x, \xi)P(d\xi)$$
$$\le p_l(\frac{t^{l+1} - \tilde{\xi}^l}{t^{l+1} - t^l}Q(x, t^l) + \frac{\tilde{\xi}^l - t^l}{t^{l+1} - t^l}Q(x, t^{l+1}))$$

$$\mathcal{Q}(x) \le \sum_{l=1}^{L} p_l(\frac{t^{l+1} - \tilde{\xi}^l}{t^{l+1} - t^l}Q(x, t^l) + \frac{\tilde{\xi}^l - t^l}{t^{l+1} - t^l}Q(x, t^{l+1}))$$

Proof: The assertion follows directly from convexity.   □

Let us now turn our attention to the multidimensional case and assume that the random variables $\xi_1, \ldots, \xi_r$ are stochastically independent.

Let us consider a partition $\Xi_1, \ldots, \Xi_L$ of Ξ consisting of subintervals

$$\Xi_l = \prod_{i=1}^{r}[a_{i,l}, b_{i,l}], \ \forall l$$

Proposition 2.12 *(Edmundson-Madansky inequality)*

$$\int_{\Xi_l} Q(x,\xi)P(d\xi) \leq \hat{Q}_l(x) \stackrel{\text{def}}{=} p_l \sum_{v \in V_l} \left[\prod_{i=1}^{r} (1 - \frac{|\tilde{\xi}_i^l - v_i|}{b_{i,l} - a_{i,l}}) \right] Q(x,v) \ \forall l. \tag{2.32}$$

$$Q(x) \leq \hat{Q}(x) \stackrel{\text{def}}{=} \sum_{l=1}^{L} \hat{Q}_l(x) \tag{2.33}$$

where V_l is the set of vertices of Ξ_l.

Proof: The assertion follows from the one-dimensional case by successive integration and induction, see Kall and Stoyan [84], Birge and Wets [12]. □

Let us introduce the denotation

$$\hat{p}_{l,v} = p_l \cdot \prod_{i=1}^{r} (1 - \frac{|\tilde{\xi}_i^l - v_i|}{b_{i,l} - a_{i,l}});$$

then the inequality (2.33) can be formulated as follows:

$$Q(x) \leq \sum_{l=1}^{L} \sum_{v \in V_l} \hat{p}_{l,v} Q(x,v). \tag{2.34}$$

Let us consider the following problem:

$$\left. \begin{array}{rl} \min c^T x + \sum_{l=1}^{L} \sum_{v \in V_l} \hat{p}_{l,v} Q(x,v) & \\ A x & = b \\ x & \geq 0. \end{array} \right\} \tag{2.35}$$

As in the case of lower bounds, inequality (2.34) implies that the optimal objective value of (2.35) is an upper bound for the optimal objective value of our two-stage problem (2.2). (2.35) is a two-stage problem with a finite discrete distribution. The interpretation is now the following: The random variable $\xi(\omega)$ has been approximated by a discretely distributed random variable $\hat{\xi}(\omega))$ taking its values at the vertices of the subdivision. This distribution has interesting extremal properties and is intimately related to moment problems, see Dupačová [27], [28], [29] and Kall [76], [78]. The connection with the moment problem leads to new bounds, see Kall [77].

Notice that in the linear affine case equality holds in (2.33).

Proposition 2.13 *The Edmundson-Madansky upper bound resulting from a finer subdivision is at least as good as the previous one.*

Proof: Straightforward. □

Frauendorfer [37] generalized the inequality (2.33) for the case of dependent random variables; the connection of this bound with multivariate moment problems has been given by Kall [74].

The above summarized analysis leading to bounds can also be carried out for simplices instead of intervals, see Frauendorfer [39]. For other types of upper bounds see the survey of Birge and Wets [12].

Computation

As for lower bounds, (2.35) is a two-stage problem with a finite discrete distribution. In this case the number of realizations equals the overall number of vertices in the subdivision, which can be very high, even for moderately dimensioned problems. Although (2.35) can theoretically be solved by any algorithm devised for this type of problems, in practice this is typically very time-consuming or cannot be carried out at all due to the size of the problem. For a fixed x (2.34) yields an upper bound. To compute this bound, an LP corresponding to a second-stage problem is to be solved at each vertex of the subdivision.

2.3.3 Bounds for chance-constrained models

In this subsection we briefly summarize bounds on the distribution function of a multivariate random variable. For a detailed presentation see Kwerel [94], Prékopa [127], [128], [130] or Kall and Wallace [85]. In connection with the Boole-Bonferroni inequalities see also Szántai [150], [152].

Let ξ be a random vector $\xi : \Omega \mapsto \mathbb{R}^r$ on the probability space $(\Omega, \mathcal{F}, P)$ with a distribution function $F(x_1, \ldots, x_r)$. Let us consider the following events:

$$A_i = \{\, \omega \mid \xi_i(\omega) < x_i \,\} \text{ and } B_i = A_i^c = \{\, \omega \mid \xi_i(\omega) \geq x_i \,\}, \quad i = 1, \ldots, r$$

then we obviously have

$$F(x) = P(\cap_{i=1}^r A_i) = 1 - P(\cup_{i=1}^r B_i)).$$

Let us introduce a counter random variable $\nu : \Omega \mapsto \{0, \ldots, r\}$ as the number of events which occur out of $B_1, \ldots, B_r$. Then we evidently have:

$$F(x) = 1 - P(\{\omega \mid \nu(\omega) \geq 1)$$

hence it is sufficient to bound $P(\{\omega \mid \nu(\omega) \geq 1)$. Let us introduce the k'th binomial moment of ν as

$$S_{0,r} = 1 \text{ and } S_{k,r} = \sum_{1 \leq i_1 \leq \ldots \leq i_k \leq r} P(B_{i_1} \cap \ldots \cap B_{i_k}) \text{ for } k \geq 1. \tag{2.36}$$

It is not difficult to show, see e.g. Takács [153], that

$$S_{k,r} = E\left[\binom{\nu}{k}\right] = \sum_{i=0}^r \binom{i}{k} \cdot P(\nu(\omega) = i) \tag{2.37}$$

holds. According to the well-known exclusion-inclusion formula of probability theory, see e.g. Rényi [135] we have:

$$P(\nu \geq 1) = S_{1,r} - S_{2,r} + \ldots + (-1)^{r-1} S_{r,r}$$

which suffers however from the combinatorial explosion because it involves all subsets of $\{0, \ldots, r\}$ not to speak about the difficulty of computing all binomial moments. Another possibility is to consider (2.37). According to these relations the unique solution of the following system of linear equations

$$\sum_{i=0}^{r} \binom{i}{k} v_i = S_{k,r}, \ k = 0, \ldots, r \tag{2.38}$$

is $v_i = P(\nu(\omega) = i)$, $i = 0, \ldots, r$ and for this solution clearly hold: $v_i \geq 0 \ \forall i$ and $P(\nu(\omega) \geq 1) = \sum_{i=1}^{r} v_i$. The idea for obtaining bounds is the following. Let us consider relaxations of (2.38), let $m \ll r$ and consider the following two LP problems (min and max over the same domain):

$$\left. \begin{array}{c} \min \ (\max) \quad \displaystyle\sum_{i=1}^{r} v_i \\[2ex] \displaystyle\sum_{i=k}^{r} \binom{i}{k} v_i = S_{k,r}, \ k = 0, \ldots, m \\[2ex] v_i \geq 0, \ k = 0, \ldots, m. \end{array} \right\} \tag{2.39}$$

Denoting the optimal objective values for the min and max problems by V_m^{min} and V_m^{max}, respectively, we obviously get the bounds:

$$V_m^{min} \leq P(\nu(\omega) \geq 1) \leq V_m^{max}.$$

Prékopa [127], [128] developed efficient algorithms for the solution of this type of problems by utilizing the structure of dual feasible bases. For generalizations see Boros and Prékopa [15]. For the special case m=2 the structural description mentioned above leads to the Boole-Bonferroni inequalities (see also Kall and Wallace [85]):

$$1 - S_{1,r} + \frac{2}{r} S_{2,r} \leq F(x) \leq 1 - \frac{2}{k+1} S_{1,r} + \frac{2}{k(k+1)} S_{2,r} \tag{2.40}$$

where $k = \left[\frac{2S_{2,r}}{S_{1,r}} \right] + 1$ with $[\cdot]$ denoting integer part.

Computation

For computing the Boole-Bonferroni bounds (2.40) we have to compute the binomial moments $S_{1,r}, S_{2,r}$. Considering their definition in (2.36), this computation just involves the one- and two-dimensional marginal distributions. In the case of a multinormal distribution these marginal distributions are again normal and for their computation efficient procedures are available.

Chapter 3

Stochastic linear programming algorithms

This section is devoted to solution approaches in stochastic linear programming. We do not intend to give a complete survey. Our aim is to summarize some of the main approaches with a detailed presentation only for those methods, which participate in the comparison.

3.1 Algorithms for two stage models

The two-stage model (2.1) is a numerically hard problem because of the second term in the objective. This term involves an integral of a function where the computation of the function-values involve the solution of an LP problem. From the theoretical point of view (2.1) is a convex nonlinear programming problem (see Section 2.1) which suggests the application of general NLP algorithms. In fact for moderately sized complete-recourse problems and for simple recourse problems Nazareth and Wets [118] proposed algorithms of this type. By utilizing the separability, convexity and piecewise linearity, Prékopa [129] and Wets [158] developed algorithms for the simple recourse model with only the RHS being stochastic. In this work we concentrate on the following approaches: Solving the algebraic equivalent LP in the discretely distributed case; discrete and stochastic approximation, respectively, for arbitrary distributions (with bounded support). discrete approximation methods and stochastic algorithms. For overviews on algorithms for two stage problems see Infanger [68], Kall [71], [73], [75], Kall, Ruszczyński and Frauendorfer [83], Kall and Wallace [85], Wets [159].

3.1.1 Algorithms for solving an equivalent LP

The algorithms considered in this subsection are designed to solve two-stage problems with a finite discrete probability distribution. Such problems can equivalently be formulated as linear programming problems (2.8) having a dual block-angular structure as shown in Figure 2.1. The algorithms considered here are essentaially algorithms for solving LP-problems with a special structure. In addition to the general surveys on algorithms see also Wets [160] for this type of methods.

The main algorithms in this class are the following:

- The L-shaped method of Van Slyke and Wets [156] presented in Section 1.2.5 as Algorithm 1.17.

- The multicut method of Birge and Louveaux [11] which is an L-shaped method with disaggregate cuts and is to be found in the same Section as Algorithm 1.18.

- The basis-reduction method of Strazicky [149] in which LP basis-reduction techniques are applied to the block–angular dual problem (2.10).

- The regularized decomposition method of Ruszczyński [138]. The algoritm is discussed in detail in Sections 1.3.3 and 1.3.4.

- New algorithms based on *interior point methods* developed recently [1] : The cutting plane method from analytic centers of Bahn, du Merle, Goffin and Vial [2] and the factorization method of Birge and Holmes [10]. Interesting proposals and discussions related to this approach can be found in Lustig, Mulvey and Carpenter [99] and Ruszczyński [140].

In the algorithms having their roots in Benders decomposition, at each iteration typically many LP's are to be solved corresponding to the realizations of the second-stage problem. Special LP-techniques, called *bunching, sifting and trickling down* have been developed to utilize the similarity of these problems, especially for the case when only the RHS is stochastic. We do not discuss them here; for these methods see e.g. Kall and Wallace [85] and Wets [159].

3.1.2   Discrete approximation methods

In this subsection we consider approximation methods which are based on successive discrete approximations of the probability distribution; for an overview see Kall, Ruszczyński and Frauendorfer [83]. For other types of approximation methods see Birge and Wets [12].

Complete recourse problems

The discrete approximation procedure for solving complete recourse problems was first proposed for $q(\xi) \equiv q$ and theoretically studied by Kall [70]. Discrete approximation algorithms based on the use of the lower and upper bounds discussed in Section 2.3 have been developed and studied by Frauendorfer [37], Frauendorfer and Kall [38] and Kall

---

[1]A new IPM version, well suited for dual block–angular problems has been developed by Mészáros, [112]. His solver BPMPD has recently been connected to SLP–IOR.

and Stoyan [84].

Below we present the method for the independent case. In the dependent case just Step 2 changes; the generalized Edmundson-Madansky inequality is to be employed. For the denotations see Section 2.3; let Ξ be a closed interval containing the support. According to Kall and Stoyan [84], for computing lower bounds the dual formulation (2.31) is used to have a hot-start facility in subsequent iterations.

Algorithm 3.1 *(Approximation method of Kall and Stoyan [84])*

Step 0. *Initialize.*   Let $k := 1$, $\Xi_1 = \Xi$; $L := 1$; $\tilde{\xi}^1 = \int_\Xi \xi P(d\xi)$ i.e. the expected value. Let ϵ_1 be the starting tolerance for selecting intervals for further subdivision and ϵ^* be a stopping tolerance.

Step 1. *Compute lower bound.*   According to the current subdivision set up and solve the dual LP (2.31). Let the corresponding optimal primal solution be x^k. If $k > 1$ then make a hotstart: Using the solution from the previous iteration construct a feasible solution with an objective value at least as high as the previous one, according to Kall and Stoyan [84]. Reduce this solution to a basic feasible solution by standard LP techniques, and use it for hot start.

Step 2. *Compute upper bound.*   According to (2.32) compute an upper bound by solving a second-stage LP at each of the vertices of the current subdivision with $x = x^k$.

Step 3. *Test for optimality.*   Check whether $\dfrac{\hat{Q}(x^k) - \tilde{Q}(x^k)}{1 + |\tilde{Q}(x^k)|} < \epsilon^*$ holds. If yes then accept x^k as a solution $\Longrightarrow$ **STOP**, otherwise:

Step 4. *Setup subdivision list.*   For each of the subintervals check whether $\dfrac{\hat{Q}_l(x^k) - \tilde{Q}_l(x^k)}{1 + |\tilde{Q}(x^k)|} \geq \epsilon_k$ holds. Set up a subdivision list consisting of those intervals, where for the deviation the above inequality holds.

Step 5. *Update subdivision tolerance.*   If the subdivision list is empty, then set $\epsilon_k := 0.5 \cdot \epsilon_k$ and **Go to Step 4** else set $\epsilon_{k+1} := \epsilon_k$.

Step 6. *Perform subdivision.*   For each of the selected intervals in turn do:

- According to refining rules select a coordinate direction across which the subinterval will be subdivided. Subdivide the interval by using a hyperlane orthogonal to the coordinate direction and passing through the conditional expectation point. Compute conditional expectations and probabilities for the two new subintervals.

Set $k := k + 1$ and **Goto Step 1**.

□

Convergence behavior

Assume that the maximum of the probabilities of the subintervals converges to zero. Under suitable further assumptions pointwise convergence and epi-convergence of the approximating sequence of expected-value functions can be proven; see Birge and Wets [12] and Kall, Ruszczyński and Frauendorfer [83], and the references therein.

For the case of independent random variables a detailed algorithm including also heuristic partition-refinement rules can be found in Frauendorfer and Kall [38]. Frauendorfer [37] published an algorithm for the dependent case. Let us remark, that the algorithm for the dependent case is the same as for the independent case, with a single exception: The upper bound is to be computed according to the generalized Edmundson-Madansky inequality.

Let us discuss the refinement rules in a more detailed form. Let Ξ_l, $l = 1, \ldots, L$ be a partition of Ξ:

$$\Xi = \prod_{i=1}^{r} [a_i, b_i] \text{ and } \Xi = \bigcup_{l=1}^{L} \Xi_l; \ \Xi_l \cap \Xi_k = \emptyset \ (l \neq k)$$

Assume that with x fixed, for this partition we have computed the lower and upper bounds according to (2.26) and (2.34), respectively. The computation of these bounds supplies also the following quantities:

- For each subinterval the conditional expected value $\tilde{\xi}^l$,

- For each subinterval a lower bound $\tilde{Q}_l(x)$ and an upper bound $\hat{Q}_l(x)$,

- For each vertex $v \in V_l$, for the second-stage problem an optimal primal solution y_v, an optimal dual solution π_v and an optimal objective value $Q(x, v)$.

For refining the partition we have to decide which of the subintervals should be further subdivided. We have seen in Section 2.3 that in the case of linearity of $Q(x, \cdot)$ on Ξ_l the lower and upper bounds are equal, a subdivision would give no improvement in the approximation. We will select those subintervals, where some measure for difference exceeds some tolerance. Let us denote by $\Delta_l(x)$ the relative difference:

$$\Delta_l(x) = \frac{\hat{Q}_l(x) - \tilde{Q}_l(x)}{1 + |\tilde{Q}_l(x)|}$$

The following empirical measures are used:

$$\Delta_l(x) \text{ or } \Delta_l(x) \cdot P(\Xi_l) \text{ or } \Delta_l(x) \cdot P(\Xi_l) \cdot 2^{\mu_l} \tag{3.1}$$

where μ_l is the number of subdivisions which lead to the specific interval. The third measure was proposed by H. Gassmann .

Having selected subintervals for subdivision the next question for each of these subintervals is, which should be the coordinate-direction across which the corresponding interval will be subdivided. Two heuristic proposals in Frauendorfer and Kall [38] are as follows. In both cases an empirical measure is defined on the edges which measures in a heuristic sense deviations from the linearity, or "degree of nonlinearity". That direction will be selected, which corresponds to an edge where the measure of nonlinearity is maximal over all edges. Let us consider a fixed subinterval and two of its adjacent vertices, $u, v \in V_l$. Denote the measure of nonlinearity for the corresponding edge by $\Lambda_{u,v}$.

- **Nonlinearity measure 1.** If the optimal dual solutions coincide at each vertex then $Q(x, \cdot)$ is linear on Ξ_l. If this is true for all adjacent vertices corresponding to a coordinate direction then $Q(x, \cdot)$ is linear on all edges along this direction; this coordinate direction can reasonably be excluded. These considerations suggest the measure $\Lambda_{u,v} = \| \pi_u - \pi_v \|$.

- **Nonlinearity measure 2.** Convexity of $Q(x, \cdot)$ and the linearity of relations (2.3) lead to the following subgradient inequalities:

$$Q(x, u) + \pi_u^T((h(v) - T(v)x) - (h(u) - T(u)x)) \leq Q(x, v)$$
$$Q(x, v) + \pi_v^T((h(u) - T(u)x) - (h(v) - T(v)x)) \leq Q(x, u)$$

and choose the minimum of the two linearization errors as our measure of nonlinearity (for the heuristic background for the minimum see Frauendorfer and Kall [38]). This gives the following measure:

$$\Lambda_{u,v} = \min\{Q(x, v) - \pi_u^T(h(v) - T(v)x, \ Q(x, u) - \pi_v^T(h(u) - T(u)x\}$$

For further heuristic rules see also Kall and Wallace [85].

For continuous distributions the computation of conditional expectations and probabilities involves (numerical) integration. To avoid this, a possible approach is to discretize the distributions prior to starting up the above algorithm. Frauendorfer and Kall [38] give a procedure for achieving ϵ-optimality w.r. to the original distribution, which relies on restart of the algorithm with increased sample sizes.

Frauendorfer [39] developed a discrete approximation method along the lines of the previous algorithm, where the support is covered by, and the subdivision is performed on Cartesian products of simplices.

Simple recourse problems

In the following we will discuss how the Kall-Stoyan algorithm specializes for simple recourse problems with only the right hand side being stochastic. We assume again that $\Xi = \prod_{k=1}^r [a_k, b_k]$ contains the support. Let us begin by summarizing the main points.

- According to (2.17) the recourse term in the objective is separable in terms of the variables χ_k. This implies that the subdivision process can be carried out separately on the intervals $[a_k, b_k]$, $k = 1, \ldots, r$.

- In the algorithm for complete recourse an upper bound was computed for a fixed x. In the simple recourse case for fixed x the objective value can be computed explicitly according to the formulas (2.17)-(2.19).

- According to (2.16) the recourse function terms $Q_k(\chi_k, \cdot)$ are linear in those subintervals of $[a_k, b_k]$ which do not contain χ_k.

- Formula (2.16) also shows that in the case $\chi_k \in I_{k,l}$ the best subdivision position is χ_k. In fact in this case $Q_k(\chi_k, \cdot)$ is linear on both of the resulting subintervals of $I_{k,l}$. This implies that for the same x the lower and upper bounds on both subintervals will be equal.

Let us assume that we have a subdivision of the intervals as follows:

$$[a_k, b_k] = \cup_{l=1}^{N_k} I_{k,l}, \; I_{k,l} \cap I_{k,t} = \emptyset, \; l \neq t; \text{ and } I_{k,l} \text{ is an interval } \forall l$$

and let us denote by $p_{k,l}$ and $h_{k,l}$ the conditional probability and conditional expectation w.r. to the k'th one-dimensional marginal distribution corresponding to the subinterval $I_{k,l}$, respectively. Let us denote the k'th row vector of T by $\hat{t}_k$. With these denotations the dual of the equivalent LP for computing the lower bound (2.31) has in the simple recourse case the following form:

$$\left. \begin{aligned} \max [b^{\mathrm{T}} u &+ \sum_{k,l} h_{k,l} v_{k,l}] \\ A^{\mathrm{T}} u &+ \sum_{k,l} \hat{t}_k^{\mathrm{T}} v_{k,l} \leq c, \\ -p_{k,l} q_k^- &\leq \quad v_{k,l} \quad \leq p_{k,l} q_k^+, \; \forall k, l. \end{aligned} \right\} \tag{3.2}$$

Now we are prepared to present the algorithm for the simple-recourse case:

Algorithm 3.2 *(Approximation method of Kall and Stoyan [84])*

Step 0. *Initialize.* Let $k := 1$, for each coordinate in turn compute the expected values. Let ϵ_1 be the starting tolerance for selecting intervals for further subdivision and ϵ^* be a stopping tolerance.

Step 1. *Compute lower bound.* According to the current subdivision set up and solve the dual LP (3.2). Let the corresponding optimal primal solution be x^k and compute χ. If $k > 1$ then make a hotstart; the Kall-Stoyan hotstart [84] now specializes as: Using the solution from the previous iteration we can construct a basis and a basic feasible solution whith an objective value at least as high as before.

Step 2. *Compute objective value.* According to (2.17) compute the exact objective value at $x = x^k$. Let us use the same denotation for it, as in the general complete recourse case for upper bounds.

Step 3. *Test for optimality.* Check whether $\dfrac{\hat{Q}(x^k) - \tilde{Q}(x^k)}{1 + |\tilde{Q}(x^k)|} < \epsilon^*$ holds. If yes then accept x^k as solution $\Longrightarrow$ **STOP**, otherwise:

Step 4. *Setup subdivision list.* For each of coordinates $k = 1, \ldots, r$ in turn select that subinterval, which contains χ_k. Now this list is nonempty, because otherwise $Q_k(\chi^k, \cdot)$ would be linear on each subinterval $\forall k$, and so the algorithm would have been stopped at the previous iteration.
For each of the subintervals (k,l) on the list check whether
$$\frac{\hat{Q}_{k,l}(x^k) - \tilde{Q}_{k,l}(x^k)}{1 + |\tilde{Q}(x^k)|} \geq \epsilon_k$$ holds, and set up a new list containing those where the inequality holds. Here the double indices refer to the corresponding terms in the sum expressing the objective- or lower bound values.

Step 5. *Update subdivision tolerance.* If the subdivision list is empty, then set $\epsilon_k := 0.5 \cdot \epsilon_k$ and **Go to Step 4** else set $\epsilon_{k+1} := \epsilon_k$.

Step 6. *Perform subdivision.* For each of the selected one-dimensional intervals in turn perform a subdivision at χ_k and compute the conditional expectations and probabilities for the two new subintervals. Set $k := k+1$ and **Goto Step 1.**

$\square$

3.1.3   Stochastic algorithms

Stochastic algorithms can be characterized by the property that now we only have convergence in the probabilistic sense; for the procedures considered here we have convergence with probability 1 (w.p. 1). The two main subclasses consist of the stochastic quasigradient algorithms and the stochastic versions of the Benders decomposition method.

Stochastic quasigradient methods were developed by Ermoliev [32], [33] and Gaivoronski [43] and implemented by Gaivoronski [44]. Ruszczyński [139] improved the method by taking into account averaged stochastic subgradients from previous steps, and by determining the step direction by solving a quadratic direction-finding problem. Marti [103] and Marti and Fuchs [104], [105] developed semi-stochastic procedures; for certain classes of distributions and at certain iteration points they replace the stochastic quasigradient by a deterministic descent direction.

Considering stochastic versions of the Benders decomposition, the method of Dantzig and Glynn [20] and Infanger [68] is based on importance sampling. We discuss here the second method of this class, the stochastic decomposition method of Higle and Sen [57], [62].

Stochastic decomposition

We assume that $q(\xi(\omega)) \equiv q$ holds.

We consider the two-stage model (2.1) under the assumptions in Section 2.1 with the following additional requirements:

- X and $\{u \mid W^T u \le q\}$ are compact sets.

- Ξ is compact.

- For all $x \in X$ the inequality $Q(x, \xi(\omega)) \ge 0$ holds w.p. 1.

The general idea of the stochastic decomposition (SD) method of Higle and Sen [57], [62] is as follows:

Let $\tilde{\xi}^1(\omega), \tilde{\xi}^2(\omega), \ldots$ be an infinite sequence of independent identically distributed random variables each of them having the same distribution as $\xi(\omega)$. Then the following probabilistic approximation is utilized:

$$Q(x) = EQ(x, \xi(\omega)) \approx S^k(x, \omega) \overset{\text{def}}{=} \frac{1}{k} \sum_{t=1}^{k} Q(x, \tilde{\xi}^t(\omega)), \ k = 1, 2, \ldots$$

According to Kolmogorov's strong law of large numbers (see e.g. Bauer [3]), for any fixed $x \in X$, $S^k(x, \omega) \to Q(x)$, $(k \to \infty)$ holds w.p. 1.

The idea is to construct lower bounding functions to this probabilistic approximation. These lower bounding functions (cuts) to $S^k(x, \omega)$ are constructed according to the following inequality, see Proposition 2.5:

$$\frac{1}{k} \sum_{t=1}^{k} (h(\tilde{\xi}^t(\omega)) - T(\tilde{\xi}^t(\omega))x)^T \pi_t^k \le \frac{1}{k} \sum_{t=1}^{k} Q(x, \tilde{\xi}^t(\omega)) = S^k(x, \omega) \qquad (3.3)$$

which holds for $\forall x \in X$, $\forall \omega \in \Omega, \forall \pi_t^k, t = 1, \ldots, k$ which are dual feasible for the second stage problem. For cut functions generated in previous iterations to become lower bounding functions also for the current Monte-Carlo integral, w.p. 1, they must be updated. A possible update is suggested by the following trivial inequality:

$$\frac{1}{k+m} \sum_{t=1}^{k} (h(\tilde{\xi}^t(\omega)) - T(\tilde{\xi}^t(\omega))x)^T \pi_t^k \ \le \ \frac{1}{k+m} \sum_{t=1}^{k} Q(x, \tilde{\xi}^t(\omega))$$

$$\overset{\text{w.p. 1}}{\le} \ \frac{1}{k+m} \sum_{t=1}^{k+m} Q(x, \tilde{\xi}^t(\omega))$$

$$= \ S^{k+m}(x, \omega)$$

where Proposition 2.5 and the almost sure nonnegativity of Q has been utilized.

Let V be the set of vertices of the dual feasible domain of the second stage problem. In the SD method a sequence of realizations $\hat{\xi}^k$ of $\tilde{\xi}^k(\omega)$ (independent observations) will be generated. At iteration k the following second stage problem will be solved:

$$\left.\begin{array}{rl} \min & q^\mathrm{T} y \\ W\,y & = h(\hat{\xi}^k) - T(\hat{\xi}^k)x^k \\ y & \geq 0 \end{array}\right\} \tag{3.4}$$

where x^k is the solution of the master problem in the previous iteration. The master problem at iteration k has the following form:

$$\left.\begin{array}{rl} \min & c^\mathrm{T} x + Z^k(x) \\ x & \in X \end{array}\right\} \tag{3.5}$$

where

$$Z^k(x) = \max_{1 \leq t \leq k}[\alpha_t^k + (\beta_t^k)^\mathrm{T} x]$$

with $\alpha_t^k + (\beta_t^k)^\mathrm{T} x$ being the cut functions. Double indices are used because all previous cuts are updated at the current iteration. The master problem can be reformulated as follows:

$$\left.\begin{array}{rl} \min & c^\mathrm{T} x + \quad \eta \\ \alpha_t^k + (\beta_t^k)^\mathrm{T} x - & \eta \leq 0, \quad t = 1,\ldots,k \\ x & \in X. \end{array}\right\} \tag{3.6}$$

Let us introduce the following notation:

$$\psi_k(x) = c^\mathrm{T} x + Z^k(x).$$

We follow the presentation of Higle and Sen in [57], [62] and formulate a conceptual algorithm first.

Algorithm 3.3 *(Conceptual SD algorithm of Higle and Sen [57])*

Step 0. *Initialize.* Let $k := 0, V_0 := \emptyset, \hat{\xi}^0 := E(\xi(\omega))$. Solve the corresponding expected value problem (2.7); let the solution be x^1.

Step 1. *Generate an observation.* Set $k := k + 1$ and randomly generate a realization $\hat{\xi}^k$ of $\tilde{\xi}^k(\omega)$.

Step 2. *Solve a subproblem.* Solve the second stage problem (3.4) by the simplex method and denote the optimal dual vertex obtained by π_k^k. Set $V_k := V_{k-1} \cup \pi_k^k$.

Step 3. *Generate a new cut; update previous cuts.*

- Compute the best vertex in the set V_k, for all previous observations, now associated with the current x^k, i.e. let

$$\pi_t^k = \; argmax \; \{(h(\hat{\xi}^t) - T(\hat{\xi}^t)x^k)^{\mathrm{T}}\pi \mid \pi \in V_k\}, \quad t = 1, \ldots, k-1.$$

- Compute the new (k-th cut) as

$$\alpha_k^k + (\beta_k^k)^{\mathrm{T}}x \equiv \frac{1}{k}\sum_{t=1}^{k} \pi_t^k(h(\hat{\xi}^t) - T(\hat{\xi}^t)x).$$

- Update the previous cuts:

$$\alpha_t^k := \frac{k-1}{k}\alpha_t^{k-1} \text{ and } \beta_t^k := \frac{k-1}{k}\beta_t^{k-1}, \quad t = 1, \ldots, k-1.$$

Step 4. *Solve the new master problem.* Solve the current master problem (3.6). Let the optimal solution be x^{k+1}; **Goto Step 1**

Convergence behavior

Proposition 3.1 *The following convergence properties hold:*

- *Let $\hat{x}$ be an accumulation point of $\{x^k\}_{k=1}^{\infty}$ and $\{x^{k_n}\}_{n=1}^{\infty}$ a subsequence such that $x^{k_n} \to \hat{x}$, $(n \to \infty)$ holds then*

  - *w.p. 1*

$$\frac{1}{k_n}\sum_{t=1}^{k_n}(h(\hat{\xi}^t) - T(\hat{\xi}^t)x^{k_n})^{\mathrm{T}}\pi_t^{k_n} \to \mathcal{Q}(\hat{x}) = EQ(\hat{x}, \xi(\omega)) \qquad (3.7)$$

  - *w.p. 1 each accumulation point of the sequence*

$$(\alpha_{k_n}^{k_n}, \beta_{k_n}^{k_n})$$

    defines a support to $\mathcal{Q}(x)$ at $\hat{x}$.

- *There exists a subsequence $\{x^{k_n}\}_{n=1}^{\infty}$ such that every accumulation point of this subsequence is an optimal solution of our problem (2.1), w.p. 1.*

Proof: See Higle and Sen [57], [62].   □

The difficulty with this proposition is that it gives no hint how such a subsequence can be found. In addition, the sequence of the lower bounds in the original Benders decomposition is not available with SD, so it is not possible to find out how far the current estimate is from the optimum.

To overcome these difficulties an *incumbent* solution is introduced. Let $\bar{x}^k$ be the incumbent at iteration k. This incumbent will be selected as a "sufficiently good" solution and kept as long as no better incumbent can be found. The cut corresponding to the incumbent will be periodically updated, thus improving the approximation locally. This gives rise to the full-scale SD algorithm:

Algorithm 3.4 *(SD algorithm of Higle and Sen [57], [62])*

Step 0. *Initialize.* Let $k := 0, V_0 := \emptyset, \hat{\xi}^0 := E(\xi(\omega))$. Solve the corresponding expected value problem (2.7); let the solution be x^1. Let $i_0 := 0$ (i_k is the index of the incumbent cut); and let $0 < \lambda < 1$ be fixed.

Step 1. *Generate an observation.* Set $k := k + 1$ and randomly generate a realization $\hat{\xi}^k$ of $\tilde{\xi}^k(\omega)$.

Step 2. *Solve a subproblem.* Solve the second stage problem (3.4) by the simplex method and denote the optimal dual vertex obtained by π_k^k. Set $V_k := V_{k-1} \cup \pi_k^k$.

Step 3. *Generate a new cut; update previous cuts.*

- Compute the best vertex in the set V_k, for all previous observations now associated with the current x^k, i.e. let

$$\pi_t^k = argmax \{(h(\hat{\xi}^t) - T(\hat{\xi}^t)x^k)^\mathrm{T}\pi \mid \pi \in V_k\}, \quad t = 1, \dots, k-1$$

- Compute the new (k-th cut) as

$$\alpha_k^k + (\beta_k^k)^\mathrm{T}x \equiv \frac{1}{k}\sum_{t=1}^{k}\pi_t^k(h(\hat{\xi}^t) - T(\hat{\xi}^t)x)$$

- Update the previous cuts, except the incumbent cut: For $t \notin \{i_{k-1}, k\}$ do

$$\alpha_t^k := \frac{k-1}{k}\alpha_t^{k-1} \text{ and } \beta_t^k := \frac{k-1}{k}\beta_t^{k-1}, \quad t = 1, \dots, k-1$$

Step 4. *Update the incumbent cut.*

- Compute the best vertex in the set V_k, for all previous observations, associated with the current incumbent $\bar{x}^{k-1}$, i.e. let

$$\bar{\pi}_t = argmax \{(h(\hat{\xi}^t) - T(\hat{\xi}^t)\bar{x}^{k-1})^\mathrm{T}\pi \mid \pi \in V_k\}, \quad t = 1, \dots, k$$

- Update the incumbent cut indexed by i_{k-1}

$$\alpha_{i_{k-1}}^k + (\beta_{i_{k-1}}^k)^\mathrm{T}x \equiv \frac{1}{k}\sum_{t=1}^{k}\bar{\pi}_t(h(\hat{\xi}^t) - T(\hat{\xi}^t)x)$$

Step 5. *Test for changing the incumbent.*
If

$$\psi_k(x^k) - \psi_k(\bar{x}^{k-1}) < \lambda \cdot [\psi_{k-1}(x^k) - \psi_{k-1}(\bar{x}^{k-1})]$$

then set $\bar{x}^k := x^k, i_k := k,$
else set $\bar{x}^k := \bar{x}^{k-1}, i_k := i_{k-1}$.

Step 6. *Solve the new master problem.*   Solve the current master problem (3.6). Let the optimal solution be x^{k+1}; **Goto Step 1**

Convergence behavior

Proposition 3.2 *Let us consider the sequence of incumbents $\{\bar{x}^k\}_{k=1}^{\infty}$; w.p. 1 there exists a subsequence such that every accumulation point is a solution of our problem.*

Proof: See Higle and Sen [57], [62].   □

In the case of finitely many incumbent changes the previous Proposition solves the problem of finding a subsequence which converges to an optimal solution w.p. 1. Let us assume in the sequel that there are infinitely many incumbent changes. Let us introduce the following denotations:

$$\theta^k = \psi_{k-1}(x^k) - \psi_{k-1}(\bar{x}^{k-1})$$

$\{x^{k_n}\}_{n=1}^{\infty}$ is the subsequence where the incumbent changed

$$n \in N^* \Longleftrightarrow \theta^{k_n} \geq \frac{1}{n}\sum_{i=1}^{n}\theta^{k_i}$$

(3.8)

The following theorem gives the possibility to identify a subsequence whose accumulation points are the solutions to our problem w.p. 1.

Proposition 3.3 *The following assertions hold:*

- *N^* is an infinite set.*

- *Every accumulation point of $\{\bar{x}^{k_n}\}_{n \in N^*}$ is an optimal solution of our problem, w.p. 1.*

- *$\dfrac{1}{m}\sum_{n=1}^{m}\psi_{k_n}(\bar{x}^{k_n}) \to \psi(x^*)$ where x^* is an optimal solution.*

Proof: See Higle and Sen [57], [62].   □

Based on this proposition stopping rules can be given, see Higle and Sen [57]. For statistically based further stopping rules relying on bootstrapping see Higle and Sen [59].

One of the stopping rules proposed in Higle and Sen [57] is the following: Let

$$\gamma^k = \frac{1}{k}\sum_{i=1}^{k}\psi_i(\bar{x}^i).$$

(3.9)

The algorithm is terminated if

$$\frac{|\psi_k(\bar{x}^k) - \gamma^{k-1}|}{\gamma^{k-1}} < \epsilon$$

(3.10)

provided that the incumbent has not been changed for a sufficiently long period of subsequent iterations and the same is true for new dual vertices encountered.

A stochastic version of Ruszczyński's regularized decomposition has been developed by Higle and Sen [60], [62] and a stochastic decomposition method with disaggregate cuts is published by Higle, Lowe and Odio [61].

3.2   Algorithms for chance constrained models

The sources of numerical difficulties in the chance-constrained model (2.23) are the multivariate distribution functions involved in the model formulation.

For the case of finite discrete distributions Prékopa [129], [130] gave a dual type algorithm. Here we will confine ourselves to the absolutely continuous case with logconcave distributions.

In the multivariate case the computation of the distribution functions for $r > 3$ is performed by Monte-Carlo techniques, see Deák [21], Szántai [150], [152] and for a survey see Deák [22]. The computation of these integrals is time-consuming and can only be done with a relatively low precision when compared to the other parts of the model. One possibility to overcome this difficulty is to design algorithms which mainly work with cheaply computable bounds discussed in Section 2.3.3. The same problem also arises at the computation of the gradient of distribution functions: Under mild assumptions this can be performed by computing the conditional distribution function which again leads to integration.

For overviews on algorithms for chance-constrained problems see Kall [71], Kall and Wallace [85], Mayer [109] and Prékopa [126], [130].

The first algorithm for solving jointly chance-constrained problems, based on a feasible direction method of Zoutendijk , has been developed by Prékopa and Deák, see Prékopa [124] and Prékopa and Deák [131]. Further solution methods not covered in subsequent sections are the following: Logarithmic barrier methods (see Section 1.4) of Prékopa [122], Rapcsák [133], Prékopa and Kelle [132] as well as a dual method of Komáromi [93].

In the methods presented in the next subsections we confine ourselves to the case S=1, i.e. we consider models with a single joint probabilistic constraint. The reason for this is that the methods are not yet tested with several chance constraints. For similar reasons we will assume that the random RHS has a joint normal probability distribution.

In the first three subsections we will assume that an initial Slater-point is available and will discuss the problem of finding such a point in the fourth subsection.

All three methods are based on some general nonlinear programming algorithm which is modified to account for the specialities of the constraint involving the distribution function.

3.2.1 A supporting hyperplane method

The algorithm has been developed by Szántai [151]. The underlying NLP algorithm is Veinott's supporting hyperplane method discussed as Algorithm 1.5 in Section 1.2.2, with Szántai's rule for moving the Slater-point.

The stochastic constraint is handled by the following linesearch procedure for locating the intersection with the boundary of the feasible domain. The linesearch is designed in such a way, that after termination we are outside the feasible domain, such that the supporting cut does not cut off pieces of it.

Assume that we are given two points y^0 and $\hat{y}$ such that $F(y^0) > \alpha$ and $\hat{y} < \alpha$ holds; let $w = y^0 - \hat{y}$. We wish to find a λ^* such that $\alpha - 2 \cdot \epsilon < F(\hat{x} + \lambda^* w) \leq \alpha$ holds.

Assume furthermore that we have a procedure for cheaply computable lower and upper bounds $F_L(y) \leq F(y) \leq F_U(y)$, see e.g. the Boole-Bonferroni inequalities in Section 2.3.3.

Algorithm 3.5 *(Linesearch of Szántai [151])*

Step 0. *Initialize.* Perform bisection on $[0, 1]$ separately with F_L and F_U to find: λ_L and λ_U such that $F_L(\hat{y} + \lambda_L w) = \alpha$ and $F_U(\hat{y} + \lambda_U w) = \alpha$ hold. Then we obviously have $\lambda_L \leq \lambda_U$ and the bisection will be continued on this interval.

Step 1. Set $\lambda_M := 0.5 \cdot (\lambda_L + \lambda_U)$ and compute $F(\hat{y} + \lambda_M w)$.

Step 2. If $F(\hat{y} + \lambda_M w) \geq \alpha + \epsilon$ then set $\lambda_U := \lambda_M$ and **Goto Step 1.**

Step 3. If $F(\hat{y} + \lambda_M w) \leq \alpha - 2\epsilon$ then set $\lambda_L := \lambda_M$ and **Goto Step 1.**

Step 4. If $\alpha - 2\epsilon < F(\hat{y} + \lambda_M w) \leq \alpha - \epsilon$ then $\lambda^* := \lambda_M$, $\Longrightarrow$ **STOP.**

Step 5. Increase the samplesize to a high level to get rather accurate function value and recompute $F(\hat{y} + \lambda_M w)$.

- If $F(\hat{y} + \lambda_M w) > \alpha$ then set $\lambda_U := \lambda_M$, **Goto Step 1.**
- Otherwise $\lambda^* := \lambda_M$, $\Longrightarrow$ **STOP.**

□

Algorithm 3.6 *(Supporting hyperplane method, Szántai [151])*

The method is specified through Algorithm 1.5, with Szántai's rule for moving the Slater point and with the linesearch algorithm 3.5.   □

For starting up, the method requires a Slater point and a point which satisfies the deterministic constraints and is infeasible w.r. to the stochastic constraint. The latter is easy to get assuming that the LP without the stochastic constraint has an optimal solution. If this optimal solution fulfils the stochastic constraint we are done otherwise we have the required infeasible point.

3.2.2   A central cutting plane method

The algorithm has been developed by the author. The underlying NLP algorithm is the modified Elzinga-Moore method 1.30 in Section 1.5. The SLP element is the linesearch algorithm 3.5 of Szántai.

Algorithm 3.7 *(Central cutting plane method)*

The method is specified through Algorithm 1.30 with the linesearch algorithm 3.5.   □

For starting up, the method in general requires a Slater point, except when by chance the center of the first inscribed hypersphere is a Slater-point.

3.2.3   A reduced gradient method

The algorithm has been developed by the author. The underlying NLP algorithm is the feasible direction reduced gradient method 1.32 in Section 1.6. Let us make a short historical comment. In the early seventies, when developing reduced gradient type algorithms for jointly chance constrained problems with a single nonlinear constraint, the author first implemented a version of GRG, see Mayer [106]. As the returning to the feasible surface (the Newton subcycle), see Section 1.6, turned out to be numerically very hard to implement for these type of problems, the author modified the method to work exclusively with feasible points. The idea was to combine Zoutendijk's P1 method [165] with the reduced gradient idea. A test version of this method was already running on a CDC computer, when the author learned that H. Sadowski in his Thesis [143] earlier proposed and theoretically analyzed almost exactly the same algorithm for the concave case. The author published a convergence proof later on for the logconcave case, see Mayer [107].

The stochastic constraint is accounted for by a linesearch procedure. This procedure tries to keep the point inside the feasible domain. From the inside the present anti zig-zag tolerance ϵ_k is used in the bisection. Let ϵ denote the tolerance for feasibility in the stochastic constraint. This means that the constraint $F(y) \geq \alpha$ is considered as active at iteration

k if $\alpha - \epsilon < F(y^k) \le \alpha + \epsilon_k$ holds. Let furthermore ϵ_0 be a tolerance on the interval length.

As in subsection 3.2.1 we assume that we have a procedure for cheaply computable lower and upper bounds $F_L(y) \le F(y) \le F_U(y)$.

Assume that we are given a $\hat{y}$ such that $F(\hat{y}) \ge \alpha - \epsilon$ holds. Let d be the direction obtained by solving the direction finding subproblem and λ_{max} the maximal stepsize imposed by the deterministic constraints. Let furthermore $\lambda_L := 0$ and $\lambda_U := \lambda_{max}$ We are looking for

$$\lambda^* = \max_{\lambda}\{\lambda \mid F(\hat{x} + \lambda d) \ge \alpha - \epsilon,\ \lambda \in [\lambda_L, \lambda_U]\}.$$

Algorithm 3.8 *(Linesearch)*

Step 0. Compute the bounds at $\tilde{y} = \hat{y} + \lambda_U d$. If $F_L(\tilde{y}) \ge \alpha - \epsilon$ then $\lambda^* := \lambda_U \Longrightarrow$ **STOP**.

Step 1. Set $\lambda_M := 0.5 \cdot (\lambda_L + \lambda_U)$, $\tilde{y} = \hat{y} + \lambda_M d$ and compute the bounds at $\tilde{y}$.

Step 2. If $\alpha - \epsilon \le F_L(\tilde{y})$ then

- If $F_U(\tilde{y}) < \alpha + \epsilon - k$ then $\lambda^* := \lambda_M \Longrightarrow$ **STOP**.
- else set $\lambda_L := \lambda_M$ and **Goto Step 6**.

else **Goto Step 3**.

Step 3. If $F_U(\tilde{y}) < \alpha - \epsilon$ then set $\lambda_U := \lambda_M$ and **Goto Step 6**.

Step 4. Compute the function value $F(\tilde{y})$ with the default samplesize in the Monte Carlo integration.

- If $\alpha \le F(\tilde{y})$ and $\alpha + \epsilon_k \le F_U(\tilde{y})$ then set $\lambda_L := \lambda_M$ and **Goto Step 6**,
- else if $\alpha \le F(\tilde{y})$ and $F_U(\tilde{y}) < \alpha + \epsilon_k$ then set $\lambda^* := \lambda_M \Longrightarrow$ **STOP**,
- else **Goto Step 5**.

Step 5. Recompute the function value $F(\tilde{y})$ with an increased samplesize in the Monte Carlo integration.

- If $\alpha + \epsilon_k \le F(\tilde{y})$ then set $\lambda_L := \lambda_M$ and **Goto Step 6**,
- else if $\alpha - \epsilon \le F(\tilde{y}) < \alpha + \epsilon_k$ then set $\lambda^* := \lambda_M \Longrightarrow$ **STOP**,
- else set $\lambda_U := \lambda_M$ and **Goto Step 6**.

Step 6. If $(\lambda_U - \lambda_L)/(1 + \mid \lambda_{max} \mid) < \epsilon_0$ then set $\lambda^* := \lambda_L \Longrightarrow$ **STOP** otherwise **Goto Step 1**. $\square$

The above linesearch may end up with a point with a small infeasibility. In this case a large weight θ is put on the stochastic constraint in the direction finding subproblem (1.85) on Page 50, otherwise this weight is being reset to the default value.

Algorithm 3.9 *(Reduced gradient method, Mayer [108])*

The method is specified through Algorithm 1.32, with the linesearch algorithm 3.8 and with the rule specified above for changing the weight in the direction finding subproblem (1.85)   □

3.2.4   Finding a Slater-point

The difficulty in finding a starting point is that the gradient components of the distribution function become very small outside a relatively small domain. This implies that e.g. exterior penalty methods simply cannot start up from an arbitrary starting point because of lacking gradient information. Below we describe a heuristic method for finding a Slater point. The algorithm is based on an idea of Szántai [151]. We assume that the expected-value problem has a solution.

We consider the following LP with a parametrized RHS:

$$
\left.
\begin{aligned}
\max \hat{c}^{\mathrm{T}} x & \\
T x - y &= \mu + \lambda \cdot \sigma \\
A x &= b \\
y &\geq 0 \\
x &\geq 0
\end{aligned}
\right\}
\tag{3.11}
$$

where $\hat{c}_j = \sum_{i=1}^{m_2} \dfrac{t_{i,j}}{\sigma_i}$, μ and σ are the expected value- and standard deviation vectors, respectively, and λ is a parameter.

Algorithm 3.10 *(Looking for a Slater point)*

Step 0. *Initialize.*   Solve the base case LP (3.11) with $\lambda = 0$ (expected-value problem). If the solution is a Slater-point, $\Longrightarrow$ **STOP**.

Step 1. *Perform parametric analysis.*   Start up a parametric analysis with increasing λ's. Continue it as long as the probability does not decrease, or the boundary of the feasibility domain is not hit. For each of the basic solutions check whether it is a Slater-point, if yes then $\Longrightarrow$ **STOP**.

Step 2. *Bisection on the last interval.*   On the last interval make bisection for the best probability value. As for the right hand sides corresponding to the bisection points the optimal basis is the same, the optimal solution can be computed by FTRAN (see Murtagh [116]).   □

In general any of the methods Discussed in Subsections 3.2.1, 3.2.2 and 3.2.3 can be used as a Phase I procedure, according to the standard NLP Phase I procedures. To start up such a procedure a point is needed, which is feasible w.r. to the deterministic constraints and the gradient components are not too small. For such a procedure the previous heuristics can be used to produce a meaningful starting point.

Chapter 4

Implementation

The subject of this chapter is implementation of SLP-algorithms. More specifically we consider implementational issues in connection with the algorithms discussed in the previous chapter. We mention that the implementation of some of the solvers was partly influenced by ongoing discussions with P. Kall .

4.1   General issues

All SLP algorithms require to solve LP problems, or perform lower-level standard LP-operations like FTRAN, BTRAN (see Murtagh [116]) or pivoting. This implies that for building LP-algorithms a modular LP-system is needed; we have based our solvers on Marsten's XMP [101]. We needed a simplex-based LP-system, as some of the algorithms explicitly utilize e.g. pivoting. According to my knowledge XMP and MINOS are the only simplex-based powerful LP-systems, which are available in source code. We have chosen XMP because it explicitly supports modular usage[1]. We considered the availability of the source as an important factor because of two reasons: On the one hand we wished to be independent of changes due to different compiler versions. On the other hand, as we need lower level LP-operations, a detailed documentation is needed; in this respect the best documentation is certainly the source itself.

As XMP is a Fortan program system, we decided to develop the solvers in Fortran.

Another common feature of SLP-solvers is that we have to work with sparse matrices. As a storage scheme for sparse matrices we use the standard Fortran scheme consisting of the following items: Two arrays containing the row indices and nonzero entries in column major order, respectively, with a third array of pointers to the beginning positions of the individual columns. An analogous scheme is used to store the realizations and probabilities for discretely distributed random variables.

---

[1] Recently MINOS has also been connected to most of the solvers.

A specific implementation of a solution algorithm will be called in the sequel a *solver*.

4.2   Two stage models

With the exception of SDECOM for stochastic decomposition, all algorithms implemented in the solvers of this section have the following common feature: At each iteration they have to solve a sequence of second-stage LP-problems, either corresponding to the realizations of discretely distributed random variables or to the vertices of a subdivision. The bunching, sifting and trickling down techniques, mentioned in Section 3.1, are not implemented in the solvers. The optimal solution and/or optimal basis corresponding to the previous LP in the sequence is used for hotstart in all of the solvers. In the comparative context this enables a fair comparison with respect to the underlying LP-algorithms.

4.2.1   QDECOM

QDECOM is a solver for two stage problems with the random variables having a finite discrete distribution. It is an implementation of Ruszczyński's regularized decomposition method Algorithm 1.27 and has been developed by A. Ruszczyński in 1985. A thorough description of the implementation can be found in Ruszczyński [138]. For new techniques (not implemented in our 1985 version) see Ruszczyński [141] [2]. The main issue in the implementation is an efficient solution method for the regularized relaxed master having a quadratic term in the objective. This can be viewed as a constrained least squares problem; in the implementation the special structure of this problem is utilized.

Due to the feature of computing disaggregate cuts a lower bound on the storage requirement is $(n_1 + 2 \cdot L) \cdot (n_1 + 1)$ double reals, with the factor 2 coming for the need for intermediate storage of the new cuts generated. Here n_1 is the number of first-stage variables and L is the number of joint realizations.

4.2.2   DAPPROX

DAPPROX is a solver for two stage complete recourse problems with the technology matrix (T) and the RHS (h) being stochastic. The random variables are assumed to be independent, having either a finite discrete, or a uniform, normal or exponential distribution. A further development for the dependent case is planned. The solver is an implementation of the discrete approximation method of Kall and Stoyan, Algorithm 3.1. DAPPROX has been developed by the author.

---

[2] A new implementation, DECOMP, of regularized decomposition has recently been developed by Ruszczyński and Świętanowski, see [142], . Connecting DECOMP to SLP–IOR is in progress.

From the implementational point of view the main issues are the computation of the upper bounds and the implementation of the refining rules. The source of the difficulty is the combinatorial explosion; having r random variables each interval has 2^r vertices. The computation of the upper bound involves solving an LP on each of the vertices of the subdivision. This can not reasonably be done by taking the subintervals in turn and solving an LP on all vertices: On a vertex in the interior this way we would solve the same LP 2^r times. The refinement rules require access to the optimal dual solutions at the vertices and additionally access to adjacent vertices corresponding to a selected coordinate direction. To handle this situation the following structures have been implemented:

Quite naturally the subdivision history is tracked by implementing a *subdivision tree* which is a binary tree with the following information stored at the vertices: Lower and upper bounds for the objective and index and position of the subdivision. The leaves additionally contain pointers to an *interval list* which represents the current subdivision.

The elements of the interval list contain the conditional expectations and probabilities as well as pointers to their 2^r vertices in the *vertex list*.

The vertex list is used to store information corresponding to the different vertices in the current subdivision. The following information is stored with an item: The two interval endpoints (lowest and highest vertex), the current optimal objective value of the corresponding second-stage LP and a pointer to a *dual-vertex list* pointing to the optimal dual solution of the LP. An additional m_2 dimensional array is also stored which contains the right hand side vector of the second stage problem (2.2) with the realization taken as the vertex and the RHS computed according to the affine sums (2.3). This vector is needed for checking the refinement rule according to the second nonlinearity measure, see Section 3.1.

The dual-vertex list contains all different dual solutions which came up in the process (an idea suggested by Higle and Sen [58] in connection with the stochastic decomposition method).

For computing the upper bound now it suffices to traverse the vertex list once. For the refinement rule with the second nonlinearity measure the information stored with adjacent vertices according to a selected coordinate direction is needed. To get this a 2^{r-1}-dimensional list of pointer pairs, pointing to the vertices, is generated. The perturbed χ vector is stored with the vertex whereas the optimal dual solutions are accessible through the pointer to the dual-vertex list. This way the selection can be carried out.

At each subdivision only the new vertices are appended to the vertex list and all structures are updated accordingly.

QDECOM is employed for solving the LP for the lower bound, implying that the present implementation of DAPPROX does not make use of the dual hot-start facility in Step 1 of the algorithm. Instead the previous primal optimum is submitted to QDECOM as a starting point.

For continuous distributions the conditional expectations are either computed from a formula if available otherwise by numerical quadrature. The interval probabilities are computed by employing the distribution function.

4.2.3 SRAPPROX

The scope of SRAPPROX is simple recourse models with only the RHS stochastic. The random variables are assumed to have either a multivariate finite discrete, or a univariate finite discrete, uniform, normal or exponential distribution. The solver implements the discrete approximation method of Kall and Stoyan, Algorithm 3.2. SRAPPROX has been developed by the author.

The algorithm uses in Step 1 optionally either the hot-start rule based on the previous dual solution or uses a hot-start from the previous primal solution.

4.2.4 SDECOM

SDECOM's present range consists of two stage complete recourse problems with the technology matrix (T) and the RHS (h) being stochastic. The random variables are assumed to be independent, their distribution is either a finite discrete, or a uniform, normal or exponential distribution. The underlying algorithm is the stochastic decomposition method of Higle and Sen [57], Algorithm 3.4. The solver has been developed by the author along the implementational guidelines by Higle and Sen [58].

The single major difference between these Guidelines and our approach is in the carrying out of the *argmax*-procedure in Steps 3 and 4. Higle and Sen propose a pointer-matrix for speeding up this operation. As this array may become quite big we decided to implement the argmax-operation directly on the cost of possibly slower runs.

The expected value problem is solved by the Benders decomposition method; the deterministic cuts may be kept optionally for stabilizing purposes.

The random variates for the sampling are generated for the discrete, uniform and exponential distribution by the inverse transform method whereas for the normal distribution by the Box-Müller method, see e.g. Deák [23] or Knuth [92]. We plan to include other methods, especially for the normal distribution.

4.3   Jointly chance constrained models

The solvers for jointly chance constrained problems in this section all use the programs of Szántai [152] for computing the values of the normal probability distribution function and its gradient, as well as bounds for the function values.

Let us notice an advantage of the model formulation (2.23) for the implementation of cutting plane algorithms: The nonzero part of the cuts is only m_2-dimensional.

For the cutting plane methods it is generally better to work with the dual formulations of the LP's which are to be solved at each iteration. The reason is that in this formulation the previous optimal basis can be used for hot-start. This feature is not yet implemented in the solvers.

All solvers use the same heuristic procedure 3.10 for finding a Slater-point.

4.3.1   PCSPIOR

PCSPIOR is a solver for jointly chance constrained problems with a random RHS having a multinormal distribution. The method implemented is Szántai's supporting hyperplane method 3.6, the solver has been developed by the author [3].

4.3.2   PROBALL

This solver is for the same type of problems as the previous one. The method which has been implemented is the central cutting plane method 1.30 discussed on Page 44; the implementation has been performed by the author.

4.3.3   PROCON

The scope of the solver is identical with the scope of the previous two solvers. The method is the reduced gradient method for jointly chance constrained problems of the author, implemented by the author.

The last optimal basis and optimal solution of the Phase0 procedure is the starting point of the procedure. Afterwards for handling the deterministic part and for carrying out the pivot-operations the LP-procedures of XMP are utilized like XFTRAN, XBTRAN, XFACT, XUPDAT. For the pivot-operations of the algorithm the analogue of partial pricing, see e.g. Murtagh [116], is implemented.

---

[3]The solver PCSP is Szántai's implementation [151] of the method. Connecting PCSP to SLP–IOR is in progress.

Chapter 5

The testing environment

In this chapter we first summarize general issues concerning empirical tests with optimization codes and summarize the characteristics of our environment. In the next section SLP–IOR is summarized which acts as a wokbench for empirical tests in this study. The third section is devoted to summarizing the main features of GENSLP, a program for randomly generating test-problems.

5.1 General aspects

The general problems concerning empirical testing of algorithms and their implementation in codes, including comparative studies involving them, have been throughly studied, also in the broader context of numerical methods, see e.g. the papers in Fosdick (ed.) [36] and Mulvey (ed.) [115]; in the field of mathematical programming see e.g. Dembo and Mulvey [25] and Crowder, Dembo and Mulvey [18]. In the empirical testing context we have to distinguish between the testing of optimization software and of optimization techniques.

The testing of optimization codes is a special case of testing software in general. In the comparative context the problem can be viewed as a multi-criteria decision problem, see Lootsma [95]. Considering optimization codes, in the field of nonlinear programming many somparative studies have been made involving a great variety of codes and algorithms. The most thorough and extensive comparative study concerning nonlinear optimization codes has been performed by Schittkowski [144]. Schittkowski also developed heuristic rules for the code-selection problem and implemented them in a ranking procedure in his expert system EMP for nonlinear programming, see Schittkowski [145], [146]. In the field of stochastic programming the author did not find results concerning the comparative behavior of SLP-software. We would like to point out that our modeling system including the solvers is aimed for research rather than for commercial usage. Our goal is to compare solution methods.

The empirical comparison of optimization techniques is an even harder problem than the comparison of codes. An optimization method is an abstract concept. What we can compare are just implementations of methods, i.e. codes. This implies that the general guidelines for testing codes, concerning the testing environment and performance measures, apply also to the testing of algorithms. The main difference is that we must be much more careful in drawing conclusions. The point is that the efficiency of an implementation of an optimization method depends to a great extent on the techniques and tricks utilized in the implementation; as an instance see the different implementations of the simplex method. Ideally several implementations of the same method should be available . Our aim in this study is to identify general trends concerning SLP-methods and their efficiency when applied to certain classes of SLP-problems. With the exception of QDECOM all implementations were made by the same persons and we gave special care to apply the same techniques and tricks for all of the solvers.

The next problem we discuss is the performance measure to use. The number of iterations is clearly inadequate because of the great differences in the methods concerning the content of one iteration. Next we consider the number of function evaluations. For two-stage problems this would mean the number of evaluations of the expected value of the second-stage objective; we cannot use it because some of the methods do not compute it at all. An alternative could be the number of diagonal subproblems solved. This could give some hints concerning the effect of a growing size of the recourse matrix, but of course only in the context of overall performance. For jointly chance-constrained problems the number of function evaluations is an important measure. All these effects are reflected in the overall computational time, so we will use this aggregate measure for performance. In our case it makes sense because all solvers are tested under the same conditions in the same computational environment.

In nonlinear programming it can be achieved that all solvers stop when a prescribed relative accuracy in the solution is achieved. Considering stochastic programming this can be achieved for the solvers for jointly chance-constrained problems. For two stage problems the the situation is radically different. Solvers which solve an algebraic equivalent solve it exactly, i.e. they only stop when the relative accuracy is the usual LP-accuracy. For solvers based on approximation schemes the stopping rule can be the same as in the NLP case, i.e. we have full control on the accuracy during the iterations. For stochastic decomposition stopping rules only act in the statistical sense, they have nothing in common with the NLP case. This implies that in drawing conclusions we must always carefully consider in what sense we refer to the fact that solvers solved the particular problems.

Considering test-problems we will work with test-problems belonging to one of the groups specified below:

- *Real-life test problems from the literature* Collections of test problems of this type have been published by Holmes [66] and King [89].

- *Variants of real-life test problems:* These are test problems obtained from the previous class of problems by imposing random perturbations on them with the help of the *Perturb* facility of SLP–IOR (see next section). The way of pertubation will be specified with the test problems.

- *Randomly generated test problems:* They have been obtained by using GENSLP as connected to SLP–IOR, see the third section of this chapter.

Next we specify the computational environment where all tests have been carried out.

Our computational environment

- **Computer:** IBM PC/Pentium, 60MHz, 16MB storage.

- **Operating system:** DOS 6.20.

- **Language:** All solvers and the test-problem generator have been developed in Fortran 77 with some Fortran 90 features for memory allocation. All reals are DOUBLE PRECISION.

- **Compiler:** Lahey F77L-EM/32 Version 5.10.

- **Measuring time:** TIMER of the above-mentioned version of the Lahey-compiler. In all solvers it is called at the very beginning and at the end of the program, the difference of the returned values is the overall elapsed time in hundredths of seconds. The data input time and the solution time is measured in the same fashion.

- **Random number generator:** RND of the above-mentioned version of the Lahey-compiler.

- **LP solver and LP subroutines:** XMP [101].

Considering reproduceability the use of the above-mentioned random number generator is certainly a drawback (compilers may unfortunately have different random-number generators in different versions) in the following sense: The random test-problem generation phase can only be reproduced when either using the same version of SLP–IOR, or else using the above mentioned compiler version for compiling GENSLP. This does not affect the tests themselves, the test problems used in this study are available with SLP–IOR on request. The SLP-solvers included into the comparison are all available themselves with SLP–IOR.

5.2   SLP–IOR: The workbench for testing SLP codes

This section is devoted to give a short overview on the workbench used for obtaining the computational results in the next chapter: The workbench facilities of SLP–IOR have been utilized for evaluating the SLP techniques. SLP–IOR is a model management system for stochastic linear programming; the design and development of SLP–IOR is a joint work of P. Kall and the author, see Kall and Mayer [79], [80], [81], [82]. The system is not closed; we work continually on the development of the model management part as well as on our solvers. Here SLP–IOR will mainly be considered from the workbench point of view. For the goals of SLP-IOR and for the model management aspects see our papers

cited above. For the subject of model management in operations resarch in general see
e.g. Dolk [26] and Bharadwaj, Choobineh, Lo and Shetty [6].

The main design principle of SLP–IOR was to build it in a close connection with a
general algebraic modeling system; we chosed GAMS for our purposes (see Bisschop
and Meeraus [13] and Brooke, Kendrick and Meeraus [17]). Besides model management
aspects discussed in our papers cited above, as a direct advantage of this approach from
the workbench point of view, we have direct access to powerful general-purpose solvers
via GAMS. The solver interface of GAMS is utilized for having a uniform interface for
our added SLP-solvers. A schema of an SLP solver-run can be seen in Figure 5.1

SLP–IOR writes the model into GAMS-format

⇓

GAMS processes the model and outputs data

⇓

A preprocessor converts data into solver input-format

⇓

Solver run

⇓

A postprocessor converts results into GAMS-format

⇓

GAMS reads the results and produces a listing

⇓

SLP–IOR reads the results

Figure 5.1: General scheme of a solver run

This general scheme is only partially valid: As GAMS is not able to represent random
variables, distribution data are being sent to some of the solvers directly by SLP-IOR in
the SLP–IOR format.

From the software-development point of view a major design decision was to develop the
system in an object-oriented style. Just for fixing terminology let us shortly review the
main aspects of the object oriented approach, for a detailed presentation see e.g. Graham
[51] and Stefik and Bobrow [148]. A **class** is a data structure defined in terms of a list
consisting of variables and procedures acting on these variables. The procedures of a class
are called **methods** of that class. If the type of some of the variables is a type defined by
a class, or a reference (pointer) to a class, then the class is called composite. An **object** is
an instance of a class, i.e. it is a variable having the type defined by a class. A **message**

is a call to a (public) method of a class, i.e. a call to a procedure which is made public in the definition of the class. The list of all messages of a class is called the (communication) **protocol** of that class. **Encapsulation** means that all procedures acting on the variables are methods of the class. The datafields (variables) are only accessible through messages; the specific implementation of data-structures and procedures is hidden, it is encapsulated into the class. Classes can be specified in **class hierarchies**. **Inheritance** means that classes inherit the data structures as well as the communication protocol of all predecessor classes in the hierarchy. We exclude **multiple inheritance**, i.e. we will only consider class-hierarchies specified by rooted trees. Methods corresponding to inherited messages can be **overridden** in the class, according e.g. to a different implementation of the data structures. **Polymorphism or late binding** means that messages are only linked at run-time to procedures thus ensuring that in the case of overridden methods that method should be carried out which corresponds to the class-type of the object which received that message. Inheritance and polymorphism facilitate the extension of the existing class-hierarchy by new (more specialized) classes. Finally let us mention that classes in the object oriented sense have many similarities to **frames** used in artificial intelligence for knowledge representation, see e.g. Fikes and Kehler [35]. Some of the main differences are that frames are usually not encapsulated and that a frame inherits the value of a **slot** whereas an object just inherits the (typed) datafield, see Graham [51].

Model classes

Next we will give a short summary on the main class hierarchies as implemented in SLP–IOR.

The various SLP model types as discussed in Chapter 2 are implemented as a class-hierarchy, reflecting the hierarchical relationships in Chapter 2. All of them have a common predecessor *MP_Model* representing a generic mathematical programming model. This is the point where extensions involving other general models, e.g. nonlinear SLP-models can be included as classes. Extending the hierarchy by more specialized models is an easy task due to inheritance and polymorphism. In fact, including e.g. simple integer recourse has been performed when the other classes already "lived" , with little effort. The model classes have a composite nature, they include as variables references to objects of the following type: *Algebraic_Structure*, *Random_Vars_Structure* and *Regression_Structure*. These three underlying structures serve for building blocks of models.

The class *Algebraic_Structure* serves for representing the underlying algebraic structure of the SLP-models which roughly consists of the various arrays and algebraic relations in the model formulations. The class contains references (pointers) to classes in the *Matrix* hierarchy and to a class *IdManager*. This latter class serves for handling names of variables and relations as well as relation types.

The *Random_Vars_Structure* represents the stochastic dependency structure and probability distributions of the random variables in the model. Among the variables there are references to classes in the *Distribution* hierarchy.

The class *Regression_Structure* represents the connection between the algebraic- and random-variable structures. It is modeled by the relations (2.3) and contains references to the classes in the *Matrix* hierarchy.

The *Matrix* class hierarchy consists of the following classes: *MatrixRoot* and *Matrix* are generic classes; *DenseMatrix* serves for arrays where all entries are stored; *Sparse_Matrix* is for matrices where only nonzeros are stored and manipulated and *TriangMatr* is a subclass of the former serving for symmetric sparse matrices.

The *Distribution* class hierarchy serves for modeling probability distributions. The hierarchy structure corresponds to the natural hierarchy of distributions in statistics. We have included most of the commonly used distributions (7 discrete and 16 continuous distributions).

Considering the communication protocols, common messages for all classes discussed above (besides the standard constructors and destructors) are the following: *Load/Save* and *ReadT/WriteT* serve for binary and ASCII I/O, respectively. *Write_to_GAMSDbase* is for writing the corresponding item in the GAMS language. *CopyYourself* builds up a copy of the object and returns the pointer to this copy. *Return_Signature* returns type-signature information which facilitates building links and serves for verification purposes. The signature is coded in 16 bytes; the code is not compressed for facilitating extensions of the hierarchies. *Receive_MDimensions/Return_MDimensions* serves for dimensioning purposes; whereas the latter is just for encapsulation, the former has in most cases memory allocation implications. Except of dimensions *Clear* resets the datafields to their default values. *Complete* and *Consistent* are Boolean functions. *Complete* returns true if and only if on the basis of the present values in the datafields the object can be considered as fully specified, e.g. the dimensions of a *Matrix* object are positive. *Consistent* returns true if and only if the present values in the datafields do not contain a contradiction, e.g. the correlation matrix of a nondegenerate multivariate normal distribution is positive definite.

The communication protocols of the classes are quite extensive. Below we just list some of the characteristic specific messages of the classes.

Some of the specific messages for the model classes are the following: *Process_Dataset* offers the datafields for changing to the user by invoking various menus and editors or just shows some graphical views like the nonzero pattern of arrays. *WriteMPS* serves for exporting the model in the S-MPS standard dataformat, see Birge, Dempster, Gassmann, Gunn, King and Wallace [9]. *Receive_LPPart/Return_LPPart* serves for injecting- or extracting the algebraic structure of a model; similar messages serve for doing the same with the other two underlying structures. *Retrieve_Results* retrieves the results after a solver run. Notice that the message *Solve* is missing. The reason is that solvers are not linked to models, the explanation for this will be given later on.

The messages for the *Matrix* class hierarchy implement besides the obvious messages for receiving-, returning- and deleting entries the usual operations in linear algebra, e.g. multiplying a matrix by a vector or transposing an array. As we usually work with large-scale arrays much care has been given to an efficient implementation of these procedures.

The messages of the *Distribution* hierarchy serve for performing operations of a statistical nature. Typical messages (with an obvious meaning) are: *Compute_Mean, Compute_SDev, DensityF, DistributionF, Quantile, RandomDeviate, ChooseSamplingMethod, Sample.*

The solver class

An important issue for model management systems is to facilitate the connection of new solvers to the system. In the workbench context this is even more important, because this way the user can perform comparative tests with her/his solver against the solvers available in SLP–IOR. Another important issue in model management systems is that the model representation and solvers should be independent. To achieve these goals we implemented the model classes and the solver class in an independent fashion, without any direct connection between them.

The specific solvers connected to SLP–IOR are all instances of the *Solver* class. Let us begin by considering the variable attributes (datafields) of the *Solver* class. Besides the obvious identification information (Name of solver, developer, etc.) the main attributes are as follows. The model-type information for defining the capability range; for each model-type in the capability range: dimensionality limitations, stochastic parts allowed, the dependency structure and distributions of random variables. Input dataformat of solver (S–MPS or SLP–IOR); dataformat of outputted results (MPS or SLP–IOR); termination codes; log- and error- files to be included into the listing. Solver control information: Run-time parameters to be asked from the user before solver startup; whether the solver is capable for hot-start and specification of possible control files.

One part of the messages in the protocol of *Solver* serves for inquiring the data discussed above from the user. A second part consists of messages for performing the necessary data-transformations on the model-data as outputted by GAMS into the input dataformat of the solver, and for converting solver output into a format readable for GAMS. The pre- and postprocessors, mentioned at the beginning of this section, function by just sending the appropriate messages to the *SolverManager* (see later) object which in turn activates the corresponding transformation by a message to the current solver object. From the rest of the messages let us discuss the following: *Configure* serves for configuring the solver run; the user is asked for the current values of control parameters and the appropriate solver-parameter and control files are written. The Boolean function *Appropriate* serves for solver selection. Upon receiving this message a solver object checks whether the current model instance belongs to its capability domain. The necessary information is acquired by the solver object through sending messages to the current model object.

Using these facilities a new SLP-solver can be connected to an executable copy of SLP–IOR such that afterwards it takes part in all solver-relevant operations precisely in the same way as our solvers connected to the system.

Control classes

The class *ModelManager* performs the top-level control. Its variable fields contain pointers to the current model instance, and to the current instances of the *Algebraic_Structure*, *Random_Vars_Structure* and *Regression_Structure* classes. These four items can be manipulated independently with the last three serving as building blocks for models. The main messages in the communication protocol are following: *Drive_SLP_IOR* operates the main menu and performs top-level control. *UseModelLibrary* is for I/O concerning models and the three underlying structures, it works by sending messages to a LibraryManager object. *ExportImport* performs its operations either in S-MPS or in GAMS format, for the latter only making export. *Process_Dataset* just sends the *Process_Dataset* message to the current model (see above). *Use_Solver_Library* serves for selecting a solver for the current model instance by sending messages to a SolverManager object. *TransformModel* is for transforming the current model instance into an instance of another model type (e.g. a two-stage model into a chance-constrained model); missing data are supplied by defaults. *UseBuildingBlocks* serves for extracting/injecting instances of the three underlying structures from/into the current model instance. *Analyze* serves for analysis purposes with respect of the current model; the *Analyze* message will be sent to the current model. The present analysis facilities include the computation of the reliablity of solution, the wait-and-see solution, the expected value of perfect information and the value of stochastic solution (for the notions see Birge [7], Kall and Wallace [85]). The *UseTools* message results in creating a *ToolsManager* object and sending it the DriveTools message. The *ToolsManager* class provides various modeling tools, which play and important role from the workbench point of view.

The *LibraryManager* controls I/O operations and inquiry concerning the model library. Only complete and consistent models are allowed to be stored in the library.

The *SolverManager* class performs control concerning solver selection and startup as well as registering of solvers connected to SLP–IOR. The main messages besides of those needed for bookkeeping of solvers are the following: *SelectSolver* performs solver selection. Upon receiving this message the SolverManager in turn sends the *Appropriate* message to the solver objects and sets up a list of those which returned true, i.e. of those solvers which considered themselves as "appropriate" for solving the current model. This means that *SolverManager* just returns a list of appropriate solvers and it is the user's responsibility for selecting one of them. We plan to develop methods for producing a ranked list, i.e. for suggesting solvers which would be best for solving the current model instance. The experimental part of this study intends to make one step into this direction by identifying general trends in solver capabilities. The messages *Configure*, *StartSolver*, *PreProcess* and *PostProcess* are for performing the corresponding activities with respect to the current solver.

A *ToolsManager* object is activated when receiving the message *DriveTools* which starts the top level control of tools. Here we just summarize those tools which were used in our empirical tests. The *Use_Algebraic_Tools* message provides tools related to the underlying algebraic structure. Besides analysis facilities from the workbench point of view the most important message concerning the algebraic structure is *Perturb* which provides perturbation facilities for the current algebraic structure. The *Use_RVarTools* serves for activating tools for the random-variable structure. The main tools are the following: Computing samples from the distributions in the *Distribution* class hierarchy and building empirical distributions based on these samples. The *Use_GENSLP* messages activates GENSLP, a program for randomly generating test-problems (see the next section).

Rules

A very important issue is to incorporate the knowledge concerning stochastic linear programming into SLP–IOR. In our case we have a very well defined problem class with clearly formulated unambiguous rules. Under such circumstances we decided to incorporate the rules in the form of polymorphic Boolean functions. Examples are *Complete*, *Consistent* and *Appropriate*, as discussed above. This technique resembles the artificial intelligence approach of rules within frames. The main difference is that in our case the rules are hard-wired into the code. Nevertheless some flexibility is provided by inheritance and polymorphism; rules can be modified by extending the class-hierarchy at the appropriate point by a new class where the previous rule is overridden.

User interface

The user performs all modeling actions through a menu driven interface. The interface software has also been developed in an object-oriented style.

Model representation

As discussed above models are represented internally in the form of a class-hierarchy. A formal model representation would consist of a (commented) listing of the class-definitons. As contrasted to algebraic modeling languages the user is not confronted with the formal model representation. She/he has to know the general model formulations, as given in Chapter 2, i.e. the model implicitly presents itself to the user in the matrix-vector-random variables-probability distributions form. In the absence of an explicit formal model definition, which can directly be manipulated by the user, the user interactions are guided by built-in rules, a context-dependent help and an on-line Users' Guide.

Considering again algebraic modeling languages, the matrix-vector form of models has the following obvious drawback: In algebraic modeling systems the user can formulate models in a natural way, corresponding to the specific problem she/he is modeling, without even knowing how a linear programming model is formulated in matrix–vector form. The matrix-vector form is produced by the compiler of such systems. In our case this matrix-vector form is the starting point, which seems to be a serious drawback in this respect. Let me first explain why we chosed this way of model representation and afterwards present our solution to this problem.

Deterministic LP is clearly a special case of SLP, so in a model management system for SLP we have to account also for this special case. There is a tremendous amount of results and systems for deterministic LP's see e.g. Bharadwaj [6]. As we have our main interest in stochastic models and as we simply do not have the manpower capacity to develop a further approach for LP-modeling, we decided to concentrate on the stochastic linear programming aspects. Considering present-day algebraic modeling languages, none of them has a facility for representing random variables. This implies that only some special classes of SLP-models can be represented in algebraic modeling languages, and only implicitly, by representing an algebraic equivalent. This can only be changed by extending the language and the compiler of the language in this direction. Such an activity presupposes access to the source of the compiler, which is not available in general, most algebraic modeling systems being commercial products. Recently there are some indications on the conceptual level (see Gassmann and Ireland [45], [46]) showing that the situation may change in this respect.

A possible alternative would be a hybrid system, where the algebraic parts are represented by an algebraic modeling language, e.g. GAMS, and the stochastic parts by our system, with our system also controlling the interplay, this way guaranteeing the building of a proper SLP model. In this approach our system would be some kind of an extension of GAMS. This approach has however very serious drawbacks which make it infeasible: On the one hand such a system would be highly inefficient due to heterogeneity, and on the other hand the user would be confronted with two quite different model representations which seems to be inacceptable.

In the light of the facts summarized above we decided to make the matrix-vector form as our starting point for SLP–IOR. In the present version the user need not know anything about GAMS. She/he has to have a copy of GAMS because SLP–IOR uses GAMS for interfacing the solvers, but GAMS does its work behind the scene.

The services of GAMS are however optionally available. The GAMS listings are on-line available for inspecting. On the other hand, GAMS can also be used for setting up the underlying algebraic structure as follows. By utilizing GAMS the user can formulate the underlying LP using the full power of an algebraic modeling language. By a facility of SLP–IOR the resulting LP can afterwards be converted to MPS-format and read by SLP–IOR. Subsequently via the model transformation facility a stochastic structure can be connected to it thus resulting in an SLP-model. If the user likes to do so she/he can make subsequent changes again in the GAMS format and repeat the cycle, but any operation concerning the SLP-model will only be possible within SLP–IOR, i.e. in the matrix-vector view.

Connections to external systems

Previously we already discussed the connection to GAMS, which is used as an external system in SLP–IOR. A further possibility is the following: Any solver can be connected as an *external solver*. This means that it just receives the model data in S-MPS standard format (see Birge, Dempster, Gassmann, Gunn, King and Wallace [9]), and can use it for whatever purposes they are needed. After solver termination, control simply comes back to SLP–IOR. We connected e.g. Greenberg's ANALYZE [52] in this way to the system. In this context "solver " is to be understood in a broad sense, besides being an optimization software it can e.g. be an editor or a file-handling system.

Implementation

The present version of SLP–IOR has been developed in Borland Pascal 7.0 and runs on IBM/PC 386 (or higher) machines with a coprocessor and requires at least 8MB storage. SLP–IOR is available free of charge for academic purposes.

5.3   GENSLP: The test-problem generator

In this section we give a brief description on GENSLP, a program for randomly generating SLP test problems.

The first version of GENSLP, see Keller [86] has been further developed by E. Keller, P. Kall and the author, and extended by the author with a facility for generating jointly chance-constrained test problems. Presently the range of problems generated by GENSLP is the following: Deterministic LP, random-, fixed-, complete- and simple-recourse two stage problems and jointly chance constrained problems with a single joint chance constraint and with only the RHS being stochastic, having a nondegenerate multinormal distribution.

Generating complete-recourse two-stage problems

Here we give a short summary on the generation method for complete recourse problems because this facility has been used for generating test problems in this study.

Let us first describe some subprocedures. In all of the procedures below randomly generating an entry means to generate it in a prescribed range. The first procedure serves for generating a nonsingular square matrix with a prescribed number of nonzeros. It is certainly not the best method for generating dense nonsingular matrices but proved to be quite efficient for sparse matrices. The idea is to generate a nonsingular upper-triangular matrix first, and increase the number of nonzeros afterwards by adding a row/column to another successively.

Algorithm 5.1 *(Generating a nonsingular square matrix)*

Step 0. *Fill up the diagonal.* Fill up the diagonal by randomly generated nonzero entries. If nonzero target achieved, **Goto Step 3**.

Step 1. *Introduce a nonzero upper-diagonal entry.* Randomly generate a row and a column index such that it corresponds to an upper off diagonal position with a zero entry. Put a randomly generated entry at this position into the matrix.

Step 2. *Check whether done.* If nonzero target achieved then **Goto Step 3**, else **Goto Step 1**.

Step 3. *Shake the matrix.* "Shake" the matrix by randomly exchanging rows and columns. If nonzero target achieved, **Exit**.

Step 4. *Add a row/column to another.* Randomly select two rows or columns and add a multiple of the first one to the second. The multiplier is chosen such that the result again lies in the prescribed range.

Step 5. *Check whether done.* If nonzero target achieved then **Goto Step 3**, else **Goto Step 4**.

□

The next procedure serves for generating a complete recourse matrix. It is based on the following proposition:

Proposition 5.1 *An $m \times n$ matrix W has the complete recourse property iff it has a full row rank and with B being an $m \times m$ regular submatrix the following polyhedron is nonempty $\{ y \mid W \cdot y = -B \cdot \mathbf{1},\ y \geq 0 \}$, where $\mathbf{1}$ is a vector with all components equal to 1.*

Proof: The assertion is an immediate consequence of Proposition 1.4 in Kall and Wallace [85].   □

Algorithm 5.2 *(Generating a complete recourse matrix)*

Step 0. *Compute nonsingular square part.* By utilizing Algorithm 5.1 compute a nonsingular square matrix. Let the corresponding partition be $W = (B, D)$ with B being the nonsingular matrix generated.

Step 1. *Compute RHS in Proposition 5.1* Compute $B \cdot \mathbf{1}$.

Step 2. *Starting fillup for D.* Generate in each column of D a nonzero entry and afterwards by possibly generating further entries ensure that each row in D also contains at least nonzero entry.

Step 3. *Check whether done.* If nonzero target achieved then **Goto Step 5**, else **Goto Step 4**.

Step 4. *Fill up D according to nonzero target.*   By randomly generated positions
and entries fill up D till the number of nonzeros target is achieved.

Step 5. *Ensure complete recourse property.*   Change the last nonzero entry in
each row such that the condition in Proposition 5.1 is fulfilled.

□

The next algorithm obviously generates complete recourse problems where the existence
of an optimal solution is ensured. The denotations in Chapter 2 will be used.

Algorithm 5.3 *(Generating a complete recourse model)*

Step 1. *Generate recourse part.*   By utilizing Algorithm 5.2 generate a recourse
matrix W having the complete recourse property. Generate T, h and q according to the prescribed range and sparsity, and for q ensure nonnegativity.

Step 2. *Generate first-stage constraints.*   Generate A and a candidate feasible
solution $x \geq 0$. Afterwards generate b in such a way that x becomes a feasible
solution, i.e. $Ax = b$ holds.

Step 3. *Generate first-stage objective.*   Generate the first-stage objective in such
a way that x becomes an optimal solution of the first-stage problem.

Step 4. *Generate regression terms.*   Generate the regression terms according to
prescribed range, sparsity and pattern. If the second-stage objective is stochastic, ensure that all entries in the regression terms are nonnegative.

□

Generating jointly chance-constrained problems

The following procedure obviously generates jointly chance-constrained problems with the
following properties: The model contains a single joint chance constraint with only the
RHS being stochastic and having a nondegenerate multinormal distribution. The optimal
solution is known, at the optimal solution the chance-constraint is active and a prescribed
percent of the deterministic constraints is also active. For the denotations see Chapter 2.

Algorithm 5.4 *(Generating a jointly chance-constrained problem)*

Step 1. *Generate correlations.*   Generate a positive definite correlations matrix.

Step 2. *Generate matrices A and T*   Generate matrices A and T according to
the prescribed range and sparsity such that in both matrices there aro no zero
rows.

Step 3. *Compute a direction.* Find vector u with $Tu > 1$ by using XMP. If such a vector does not exist then succesively change negative entries to positive till XMP finds a solution. Compute $v = T \cdot u$. Randomly generate a base point $x_B \in \mathbb{R}^{n_1}$ according to prescribed range.

Step 4. *Generate expected values and standard deviations.* Compute expected values according to $\mu = T \cdot x_B$ and standard deviations as a prescribed percentage of the magnitude of μ; if it would be 0, set it to 1.

Step 5. *Compute Slater-point and solution.* By utilizing the linesearch techniques for chance-constrained problems in Chapter 3 compute λ_S and λ^* such that for $y^S = \mu + \lambda_S \cdot v$ and $y^* = \mu + \lambda^* \cdot v$ the following realations hold: $F(y^*) = \alpha$ and $f(y^S) > \alpha + \alpha_S$ with α being the prescribed probability level and α_S being the Slater level. Finally compute a going-to-be Slater point and optimal solution as $x^S = x_B + \lambda_S \cdot u$ and $x^* = x_B + \lambda^* \cdot u$.

Step 6. *Compute b and fix relation type.* Compute b as $b = A \cdot x^*$ and determine the direction of the inequality componentwise such that x^S becomes feasible. According to the prescribed amount of active deterministic constraints relax the other constraints by modifying b through a randomly generated nonzero slack value.

Step 7. *Compute c.* Compute $\nabla F(T \cdot x^*)$ and determine c such that at x^* the Kuhn-Tucker conditions hold.

$\square$

5.4 Availability

The model management system SLP–IOR is a steadily growing system. P. Kall and the author continuously extend it by adding new features as their research in the field progresses. The latest version of SLP–IOR is always available free of charge for academic purposes.

A distribution copy of the system contains the current version of the model management system SLP–IOR (including also GENSLP in an integrated manner) and all of the SLP solvers used in the computational tests of the next Chapter. A User's Guide is also integrated and and can be viewed or printed in the first menu level of the system.

For using the full power of SLP–IOR the User has to have her/his own GAMS system, Version 2.25. Without having GAMS none of the solvers can be started up and the User will only be able to perform the following activities: Formulating an LP or an SLP problem; generating variants of it; analyzing the problems; randomly generating test problem batteries. Exporting the formulated problems in MPS format (for LP) or in S-MPS format (for SLP) and subsequently solving them with an (S)LP solver at the User's disposal.

The distribution copy of SLP–IOR consists of an installation program, the SLP–IOR system files and the binary solver files in a compressed form along with instructions for the installation procedure. After installation SLP–IOR occupies $\sim$ 12 MBytes on the hard disk; without the installation files this is $\sim$ 9 Mbyte.

A copy of SLP–IOR for academic purposes can be downloaded via Internet by utilizing FTP. For obtaining a copy please send an e–mail to one of the authors: P. Kall (kall@ior.unizh.ch) or the author of this book (mayer@ior.unizh.ch).

The test problem batteries used in the next Chapter can also be downloaded via Internet in the form of a compressed self–expanding file. The test problems are in the input dataformat of SLP–IOR and after expansion they occupy $\sim$ 5.6 MBytes storage on the hard disk. For obtaining a copy please send an e–mail to the author (mayer@ior.unizh.ch).

Chapter 6

Computational results

6.1 Test problems

6.1.1 Two stage test problems from the literature

In this subsection we list the test problems from the literature which were used in the study. For a detailed description see the references associated with them and for collections of test problems consult Holmes [66] and King [89].

- **AIRCRAFT:** Dantzig's aircraft scheduling problem, see Dantzig [19].

- **APL1P:** A power systems capacity expansion planning test problem of Infanger [67].

- **CEP1:** A machine capacity planning problem of Higle and Sen [60], [62].

- **PGP2:** A power systems capacity expansion planning problem of Louveaux and Smeers [96].

- **REFINE:** An illustrative refinery example from Kall and Wallace [85].

The following test problems are due to Birge [8] who has built them on the basis of deterministic LP problems from the literature. For the underlying LP problems see Ho and Loute [63] and the references therein.

- **SCAGR7:** A dairy farm expansion model.

- **SCRS8:** A technological assessment model for the transition from fossil fuel to renewable energy sources.

- **SCSD8:** A structural design optimization model.

- **SCTAP1:** A dynamic traffic assignment model.

	AIRCRAFT	APL1P	CEP1	PGP2	REFINE
type	SR	FR	FR	CR	CR
st. parts	h	h, T	h	h	h, T
m_1	4	2	9	2	1
n_1	17	2	8	4	2
m_2	5	5	7	7	2
n_2	10	9	15	16	2
r	5	5	3	3	4
nrea	750	1280	216	576	—
nrows in LP equ	3754	6402	1521	4034	—
ncols in LP equ	7517	11522	3248	9220	—
nnonz in LP equ	20267	21201	6712	18440	—

Table 6.1: First group of two-stage test problems from the literature

In all of the test problems listed above the random variables are independent and with the exception of REFINE, they have a finite discrete distribution. In REFINE there are 4 random variables; the first two have a 99% truncated normal distribution, the third has a uniform distribution whereas the fourth a 99% truncated exponential distribution. The main characteristics of the test problems are summarized in Tables 6.1 and 6.2. The denotations of Chapter 2 are used throughout. Remind that m_1 and n_1 denote the number of rows and columns, respectively, in the first stage whereas m_2 and n_2 are the analogous dimensions in the second stage. The number of random variables is denoted by r. The abbreviations used in the tables are the following: "SR" stands for simple recourse; "CR" means complete recourse; "FR" stands for incomplete fixed recourse; "st. parts" means the stochastic parts of the model; "nrea" stands for the number of joint realizations of the random variables in the discretely distributed case and "LP equ" means the algebraic equivalent LP. The abbreviations "nrows" "ncols" and "nnonz" stand for number of rows, number of columns and number of nonzero entries, respectively.

Checking of the complete recourse property has been carried out by the *Analyze* facility of SLP–IOR.

Notice that 6 out of 9 test problems do not have the complete recourse property. The algorithms underlying DAPPROX and SDECOM have been discussed in Section 3.1 under the asumption of complete recourse. Observe however that the following weaker assumption would be sufficient: It suffices that the second stage problem (2.2) on Page 53 has a solution for any x which fulfills the first stage constraints. Models having this property are called models with *relatively complete recourse*.

With the exception of AIRCRAFT, APL1P and REFINE, the test problems listed in this subsection have been downloaded from freebie.engin.umich.edu, University of Michigan, see Holmes [66]. The test problems are available there in the S-MPS standard dataformat (Birge, Dempster, Gassmann, Gunn, King and Wallace [9]).

	SCAGR7	SCRS8	SCTAP1	SCSD8
type	FR	FR	FR	FR
st. parts	h	h	h	h
m_1	15	28	30	10
n_1	20	37	48	70
m_2	38	28	60	20
n_2	40	38	96	140
r	3	3	3	3
nrea	864	512	480	432
nrows in LP equ	32847	14364	28830	8650
ncols in LP equ	34580	19443	46128	60550
nnonz in LP equ	108906	50241	169068	190210

Table 6.2: Second group of two-stage test problems from the literature

6.1.2   Variants of standard two-stage test problems

The test problems in this subsection have been obtained by modifying some of the problems listed in the previous subsection. Most of the variants were obtained by imposing random perturbations on some of the arrays in the original model. The modification has been carried out by using the *Perturb* facility of SLP–IOR.

The random seeds[1] for the sampling which supplies the random variates for the perturbation were as follows:

12317, 21579, 57913, 56321, 43161, 34671, 17181, 22192, 47811, 54312.

- **PGP2Xi,** $i = 1, \ldots, 10$: These test problems were obtained from PGP2 by imposing a perturbation on the second stage objective q, see Louveaux and Smeers [96]. The operating costs of the first operating mode were perturbed by adding standard normal variates to them; the costs for the remaining operation modes were computed on the basis of the perturbed values, as described in [96] for the original model. (The original objective coefficients are in the range $[32, 55]$).

- **CEP1Xi,** $i = 1, \ldots, 10$: In CEP1 we perturbed the first stage matrix elements corresponding to the time of maintenance (measured in hours) needed for one hour operation of a machine, see Higle and Sen [60]. The perturbation consisted in adding a normal variate with zero mean and standard deviation 0.005.

- **REFINED:** This model has been obtained from REFINE by discretization, using the *Discretize* facility of SLP–IOR (i.e. the original distribution is approximated by an empirical distribution). The number of subintervals of equal lengths was

---

[1] As SLP–IOR is developed in Borland Pascal 7 the seeds are for the random number generator function *Random* of this version of Borland Pascal. In the following integer seeds in the range $(1, 65535)$ will always refer to this random number generator.

13, 11, 7, 11 for the random variables, respectively. This way a complete recourse model arises with discretely distributed random variables. The number of joint realizations is 11011. The random variate generation for the normal distributions has been carried out by the Kinderman-Ramage method, see e.g. Kennedy and Gentle [88], and for the exponential distribution by the inverse transform method. The samplesize was 30000 and the random seed 12317.

- **REFINE1:** In this variant of REFINE the distribution of the fourth random variable, originally an exponential distribution, has been replaced by a normal distribution having the same first and second moments.

6.1.3 Randomly generated two stage test problems

The test problems described in this subsection have beeen generated by GENSLP and subsequently also processed by SLP–IOR.

The seeds[2] for each test-problem battery generated by GENSLP were the same; they are listed below. The ordering corresponds to the indexing of test problems within a family.

0.506706, 0.058624, 0.516823, 0.584069, 0.469138, 0.224567, 0.833035, 0.186907, 0.536156, 0.545647, 0.976729, 0.638727, 0.672190, 0.660166, 0.970199, 0.955168, 0.225619, 0.511337, 0.285938, 0.531746.

In order to eliminate the effect of different LP subproblem sizes on the solution time, the dimensions are kept fixed within a test problem battery. The test problem batteries are as follows:

- **CR1Ni** $i = 1, \ldots, 20$: The members of this test problem set are complete recourse problems with only the RHS being stochastic. The number of random variables is 10 and they are independent. The distributions are truncated normal distributions; the truncation level is 99%. The models were generated as follows. Complete recourse problems with the following characteristics were generated by GENSLP first:

    - $m_1 = 5$; $n_1 = 10$; $m_2 = 10$; $n_2 = 30$; $r = 10$.
    - Stochastic part: h.
    - Densities: A 40%; W 20%; h 95%; q 50%; the rest 10%.
    - Magnitudes: For all arrays the magnitude of the entries is ≤ 10.0.
    - q was generated as nonnegative; the regression terms were unit vectors (see (2.3) on Page 54); matrix A has a full row rank and W is a complete recourse matrix.

---

[2]As GENSLP has been developed in Lahey Fortran F77L-EM/32 Version 5.10 the seeds are for the random number generator functions *RANDS and RND* of this version of Lahey Fortran. In what follows real seeds in the range $(0, 1)$ will refer to this random number generator functions.

By utilizing SLP–IOR the random variables were endowed with the following random variable structure: They are independent, all have a 99% truncated normal distribution with expected value zero and with standard deviation chosen as 20% of the magnitude for nonzero RHS components and 1 for zero RHS components.

- **CR1Di,** $i = 1, \ldots, 20$: This test battery was obtained from the previous (SR1Ni) test problem set by discretizing the distributions. With the help of SLP–IOR the distribution of the random vector was discretized componentwise for each model in the battery CR1Ni as follows. First a sample was computed with samplesize 20000 and seed 43175; the method was the Kinderman-Ramage procedure see e.g. Kennedy and Gentle [88]. Afterwards the empirical distribution was built on the basis of 10 intervals of equal length (equidistant subdivision) for the random vector, componentwise.

  The resulting models are complete recourse problems with only the RHS being stochastic. The components of the RHS are independent random variables with finite discrete distributions; the number of the joint realizations is 10^{10}.

- **CR2Xi,** $i = 1, \ldots, 20$: This test battery consists of complete recourse problems with the RHS (h) and technology matrix (T) both being stochastic. The random variable structure consists of 5 independent random variables with two of them having a 99% truncated standard normal distribution and the rest a uniform distribution on $[-10, 10]$. The regression terms (see (2.3) on Page 54) were generated such that the resulting models in this class have alltogether 20 dependent random entries; 10 in the RHS and 10 in the matrix. Notice that although the first three and the last two random variables are identically distributed, the random entries in the technology matrix and RHS are far from being identically distributed.

  The generating procedure was as follows: Complete recourse problems were generated by GENSLP first, with the characteristics:

  - $m_1 = 5$; $n_1 = 12$; $m_2 = 8$; $n_2 = 16$; the number of random variables is 5.
  - Stochastic parts: h and T.
  - Densities: A 40%; T 20%; W 30%; h 100%; q 50%; the rest 10%.
  - Magnitudes: For all arrays the magnitude of the entries is ≤ 10.0.
  - The second stage objective q was generated as nonnegative.
  - The regression terms (see (2.3) on Page 54) for h have the density of 100%. The regression terms T have a nonzero pattern which is a subset of the nonzero pattern of the constant term and their density is 10%.
  - Matrix A has a full row rank and W is a complete recourse matrix.

Using SLP–IOR the random variable structure described above has been injected into the model.

- **CR2Di,** $i = 1, \ldots, 20$: The problems in this class were obtained by discretizing the previous models (CR2Xi) analogously as in the case of CR1Ni but now with 3 intervals of equal length, componentwise. The resulting models are complete recourse problems with the RHS and the technology matrix both being stochastic. The 20 random entries are dependent, the number of joint realizations is $3^5 = 243$.

- **CR2Ei,** $i = 1, \ldots, 20$: The difference to the previous class is that now a finer discretization has been made: The random variables were discretized on the basis of 5 equidistant intervals thus resulting in models with $5^5 = 3125$ joint realizations.

- **CR2Fi,** $i = 1, \ldots, 10$: This test battery was also obtained from CR2Xi by discretization; the number of subintervals of equal length is now 7, 7, 7, 5, 5 for the random variables, respectively. A building block containing the common random variable structure of the models belonging to CR2Xi has been discretized componentwise and afterwards injected into the models to obtain the battery CR2Fi. The random seed for the sampling was 12317. The random variables in these models have $7^3 \cdot 5^2 = 8575$ joint realizations.

- **SR1Ni,** $i = 1, \ldots, 20$: The members of this family are simple recourse problems with only the RHS being stochastic. The RHS is a 90-dimensional random vector with independent components. The components have 99% truncated normal distributions.

They were generated as follows. First simple recourse problems with the following characteristics were generated by GENSLP:

  - $m_1 = 200$; $n_1 = 400$; $m_2 = 90$; $n_2 = 180$; $r = 90$.
  - Stochastic part: h.
  - Densities: A, T 2%; h 90%; q 50%; the rest 10%.
  - Magnitudes: For all arrays the magnitude of entries is chosen as ≤ 10.0.
  - q is generated as nonnegative; the regression terms are unit vectors and A has a full row rank.

By utilizing SLP–IOR the random variables were chosen as independent, all having a normal distribution with mean 0. The standard deviations were chosen as 10% of the magnitude of the RHS-components for nonzero components, otherwise they were set to 1.

- **SR1Di** $i = 1, \ldots, 5$ were obtained from SR1Ni by discretizing the distribution on the basis of 10 intervals of equal length, componentwise. Sampling was made with a sample size of 20000, and the sampling method was the Kinderman-Ramage procedure see e.g. Kennedy and Gentle [88]. The seed for the sampling was 43175. The number of joint realizations is 10^{90} for these models.

	CR1Ni	CR1Di	SR1Ni	SR1Di
type	CR	CR	SR	SR
st. parts	h	h	h	h
m_1	5	5	200	200
n_1	10	10	400	400
m_2	10	10	90	90
n_2	30	30	180	180
r	10	10	90	90
nrea	—	10^{10}	—	10^{90}
nrows in LP equ	—	$\approx 10^{11}$	—	$\approx 9 \cdot 10^{91}$
ncols in LP equ	—	$\approx 3 \cdot 10^{11}$	—	$\approx 1.8 \cdot 10^{92}$
nnonz in LP equ	—	$\approx 7 \cdot 10^{11}$	—	$\approx 9 \cdot 10^{92}$

Table 6.3: Randomly generated problems with only h being stochastic

	CR2Xi	CR2Di	CR2Ei	CR2Fi
type	CR	CR	CR	CR
st. parts	h, T	h, T	h, T	h, T
m_1	5	5	5	5
n_1	12	12	12	12
m_2	8	8	8	8
n_2	16	16	16	16
r	5 (20)	5 (20)	5 (20)	5 (20)
nrea	—	243	3125	8575
nrows in LP equ	—	1949	25005	68605
ncols in LP equ	—	3900	50012	137212
nnonz in LP equ	—	14362	184401	505950

Table 6.4: Randomly generated problems, h and T are both stochastic

An overview of the characteristics of the randomly generated problems is given in Tables 6.3 and 6.4. In Table 6.3, in the row labeled by "r" the first number is the number of random variables and the number in parentheses is the number of random entries in the model which results when applying the affine sums (regression terms) (2.3) on Page 54.

6.1.4 Jointly chance constrained test problems from the literature

The following standard test problems will be used, for their dimensionalities see Table 6.5. The distribution is multinormal in all of the jointly chance constrained test problems used.

- **STABIL:** A planning model for the electric energy sector in Hungary, due to Prékopa, Ganczer, Deák and Patyi [131].

	STABIL	WATRES
m_1	48	7
n_1	46	5
m_2	4	3
r	4	3

Table 6.5: Jointly chance constrained test problems from the literature

- **WATRES:** A water resources planning model of Dupačová, Gaivoronski, Kos and Szántai [30].

In both test problems only the RHS is stochastic and has a multinormal distribution. While WATRES is, similarly to all other problems used in this study, a minimization problem, STABIL is a maximization problem. In connection with STABIL the following correlation matrix will be used throughout:

$$\begin{pmatrix} 1.0 & -0.7 & 0.4 & 0.4 \\ -0.7 & 1.0 & 0.1 & 0.1 \\ 0.4 & 0.1 & 1.0 & 0.9 \\ 0.4 & 0.1 & 0.9 & 1.0 \end{pmatrix}.$$

6.1.5   Variants of standard jointly chance constrained problems

The following variants of STABIL will be used:

STABILj, $j = 1, \ldots, 10$: These test problems were obtained from STABIL by using the *Perturb* facility of SLP–IOR. The right-hand side of the first stage has been perturbed by adding normal variates to them, with zero mean and a standard deviation being 2% of the magnitude of the RHS component. The random variates were generated in the same fashion as in the case of PGP2X and CEP1X, the random seeds were as follows:

25713, 34501, 43215, 41651, 53163, 12135, 21215, 33231, 17165, 23217.

6.1.6   Randomly generated jointly chance constrained problems

These test problems were generated by utilizing GENSLP.

J1CCi, $i = 1, \ldots, 20$: The memebers of this family of test problems have the following characteristics:

- $m_1 = 40$; $n_1 = 60$; $m_2 = 4$; $r = 4$.

- Stochastic part: h.

- Densities: A 10%; T 10%; correlation matrix 80%.

- Magnitudes: For all arrays the magnitude of the entries is ≤ 10.0.

- Matrix A has a full row rank; the correlation coefficients have an arbitrary sign.

The seeds used were as follows:

0.506706, 0.706341, 0.531976, 0.417823, 0.469138, 0.224567, 0.833035, 0.186907, 0.536156, 0.545647, 0.976729, 0.638727, 0.672190, 0.660166, 0.312765, 0.955168, 0.834331, 0.681215, 0.285938, 0.531746.

6.2   Test runs for two stage problems

The default parameter settings for the solvers were as follows:

- *DAPPROX:* The maximal number of subdivisions is 100. This is for approximately 10 random variables, for models with fewer random variables it may be set higher. The maximal number of different dual solutions is 200. The coordinate selection strategy corresponds to nonlinearity measure 2 in Section 3.1 and the interval to be subdivided is selected according to the second item in (3.1) on Page 71. The starting tolerance in Algorithm 3.1 on Page 70 is $\epsilon_1 = 0.1$. The cut-tolerance for QDECOM, used for solving the LP problems for the lower bound (see Subsection 4.2.2), is 10^{-8}.

- *QDECOM:* The tolerance for cuts is 10^{-8}, see Ruszczyński [138].

- *SDECOM:* The maximal samplesize is 1000; the maximal number of dual vertices is 200. In the description of the SD algorithm 3.4 on Page 78 we presented the method with updating the incumbent at each iteration. It is sufficient to update the incumbent according to a fixed period; the default value of this period is set to 5. A cut is considered to be loose if it is inactive in 10 subsequent iterations; loose cuts are dropped at a period of 10 iterations. The termination criterion 3.10 on Page 79 has been implemented. It takes effect if the incumbent was not changed at 20 subsequent iterations and in the last 30 iterations no new dual vertex was discovered. The stopping tolerance is $\epsilon = 10^{-3}$; the descent factor is $\lambda = 0.7$. Maximum 100 deterministic Benders cuts are kept, see Subsection 4.2.4 on Page 88. The random seed for sampling is 0.231452613. The number of joint realizations for this group is 10^{90}.

- *SRAPPROX:* The maximum number of coordinate splits is set to 500 and the stopping tolerance to 10^{-6}.

Although the default values of solver parameters may freely be changed in an SLP–IOR
session, we carefully avoided changing for efficiency reasons the default parameter settings. The computational results reported here have been obtained with the default
settings, except of parameters directly connected to memory allocation. There were some
runs, involving large-scale models, where these parameters had to be changed (remind
that the solvers are coded in Fortran). For runs with changed parameters this fact is
explicitly stated in the tests below.

The starting point of DAPPROX, SDECOM and SRAPPROX has been computed as part
of the algorithm by solving the expected-value problem (2.7). QDECOM obtained the
zero vector as a starting point; the Phase I procedure is part of the solution method. For
all four solvers the computing time of the Phase I procedure is included in the computing
times reported.

In this study all computational results involve a single solver run. We do not investigate
the issues of hot start and the combination of algorithms.

All computing times reported are in seconds. Time is denoted in the table headers as
"t" whereas objective function values by "f". Time means elapsed solution time, i.e. the
data input time is not included. Further general abbreviations: "LB" and "UB" denote
lower and upper bounds, respectively; "ItCnt" means the number of iterations; "Interv"
stands for the number of intervals in the final subdivision. The abbreviation "diag LP"
means the second stage LP problem with x fixed and "Master" stands for the master
LP problem. The abbreviation "ndiag LP" denotes the overall number of second stage
LP problems solved by DAPPROX for computing upper bounds and "qdiag LP" is the
overall number of second stage LP's solved by QDECOM.

6.2.1   Test problems from the literature

<u>TEST #1</u>

The goal of these runs is to give a first impression on the comparative behavior of
DAPPROX and QDECOM on the basis of test problems with known optimal objective values.

The test problems listed in Subsection 6.1.1 having a finite discrete distribution have
been solved by QDECOM and DAPPROX. The stopping tolerance for DAPPROX was
$\epsilon^* = 10^{-7}$. The results are shown in Table 6.6 with 9 decimals for the objective values.
Considering computing time the averages are 40.5 for QDECOM and 45.9 for DAPPROX,
respectively. The results on this test problem set show a similar average performance of
the two solvers.

For solving CEP1 with QDECOM, and consequently also for DAPPROX which uses

	QDECOM		DAPPROX		
	f	t	LB on f	UB - LB	t
AIRCRAFT	1567.04219	42.3	1567.04219	$0.2 \cdot 10^{-12}$	34.7
APL1P	2464.23206	72.3	2464.23256	$0.5 \cdot 10^{-12}$	157.8
CEP1	355159.956	11.5	355159.956	$0.2 \cdot 10^{-9}$	29.6
PGP2	447.324346	40.5	447.324328	$0.33 \cdot 10^{-4}$	37.6
SCAGR7	-834539.803	105.7	-834540.048	$0.12 \cdot 10^{-1}$	15.9
SCRS8	1074.06551	8.5	1074.06551	$0.5 \cdot 10^{-10}$	16.7
SCTAP1	284.500000	27.3	284.500000	0.0	2.2
SCSD8	25.8000000	16.2	25.8000000	$0.2 \cdot 10^{-13}$	72.2

Table 6.6: Runs on standard test problems with a discrete distribution

	ItCnt	Interv	ndiag LP	diag LP t	Master t
AIRCRAFT	16	127	18880	20.3	12.4
APL1P	39	149	89888	113.0	40.6
CEP1	28	80	9424	18.4	9.0
PGP2	32	92	11248	18.6	16.9
SCAGR7	17	25	1560	9.1	5.4
SCRS8	16	38	2456	10.4	4.8
SCTAP1	3	3	48	0.7	0.9
SCSD8	16	111	4552	37.1	33.6

Table 6.7: Some details on the DAPPROX runs

QDECOM, the cut tolerance level had to be set to 10^{-7}.

Let us notice that AIRCRAFT is a simple recourse model; the computational time is 0.3 sec when solving it with SRAPPROX.

Another comment concerns the optimal objective value of CEP1. In [66] the value 179.9998 is published which is false; the correct value is that in Table 6.6, i.e. 355159.956. The source of the error is the datafile CEP1.COR. The penalty costs (400 each) are in this file without a decimal point which causes the Fortran solvers to read them according to their specific input format. If the solver reads the field e.g. according to Fortran format F12.4, then the penalty becomes 0.4 and afterwards the solver produces the published erroneous objective value.

Some details on the DAPPROX runs are shown on Table 6.7. The most difficult problem for DAPPROX was APL1P where h and T are both stochastic.

QDECOM, as all solvers designed for solving the algebraic equivalent LP, solves it to the usual precision for LP's. Discrete approximation algorithms provide a direct way of obtaining solutions with a prescribed accuracy in the objective. Table 6.8 shows the results obtained by DAPPROX for lower accuracy levels, i.e. with the stopping tolerance

	$\epsilon^* = 0.0001$ **(0.01%)**		$\epsilon^* = 0.001$ **(0.1%)**		$\epsilon^* = 0.05$ **(5%)**	
	LB on f	*t*	*LB on f*	*t*	*LB on f*	*t*
AIRCRAFT	1567.04219	34.6	1567.04219	34.6	1557.41640	9.5
APL1P	2464.01102	56.1	2462.86817	14.22	2428.33856	1.5
CEP1	355137.695	21.9	354857.023	12.1	345318.077	5.1
PGP2	447.310819	22.0	447.238375	11.9	442.105187	4.1
SCAGR7	-834556.129	4.3	-834644.914	2.7	-834711.119	1.3
SCRS8	1074.053782	11.4	1073.05640	8.5	1047.75590	3.0
SCTAP1	284.500000	2.3	284.500000	2.2	247.500000	1.4
SCSD8	25.8000000	72.3	25.8000000	72.3	25.3650000	18.9

Table 6.8: DAPPROX runs with different accuracy levels for the objective

correspondingly set higher as given in the first header line of the table.

When comparing the computing times in Table 6.6 having a very high accuracy and those in Table 6.8, a substantial reduction in computing time can be observed due to lower required accuracy.

TEST #2

The goal of these runs is similar to the goal of the previous run with the additional aim of getting some information on the quality of the solutions supplied by SDECOM.

We have run SDECOM on the standard test problems with the exception of SCAGR7. This problem does not fulfill the nonnegativity assumption on the recourse costs which is presupposed for stochastic decomposition, see Subsection 3.1.3. The results are summarized in Table 6.9.

In the header line of this table Δ_{SD} is defined as

$$\Delta_{SD} = \frac{\| x_D - x_{SD} \|}{1 + \| x_D \|}$$

where x_D and x_{SD} are the solutions obtained by DAPPROX and SDECOM, respectively, and the norm is the Euclidean norm.

In the case of SDECOM the last incumbent is returned by the solver. A small Δ_{SD} indicates that the solution returned by SDECOM is close to an optimal solution. If Δ_{SD} has a relatively high value then no conclusion can be drawn for two reasons: On the one hand according to the knowledge of the author nothing is known about the uniqueness of the solutions in the test problem set. On the other hand the sequence of incumbents requires further analysis before returning a solution, see Proposition 3.3.

The other abbreviations in the header line are: "Obj_est" stands for the objective estimate according to (3.9) on Page 79; "Sas" is the samplesize; " Inc" is the number of incumbent solutions and " Dv" is the number of different dual vertices encountered. The run with

	Obj_est	t	Sas	Inc	Dv	Δ_{SD}
AIRCRAFT	1544.79711	22.0	142	6	31	$1.3 \cdot 10^{-2}$
APL1P	2459.42389	125.5	1000	85	9	$2.8 \cdot 10^{-1}$
CEP1	429617.716	6.5	90	3	10	0.0
PGP2	455.605406	10.4	105	7	21	$7.5 \cdot 10^{-2}$
SCRS8	1035.84651	7.4	66	7	14	$1.4 \cdot 10^{-13}$
SCTAP1	252.122714	6.8	59	1	2	$9.8 \cdot 10^{-2}$
SCSD8	24.1656833	11.4	61	5	27	$2.3 \cdot 10^{-16}$

Table 6.9: SDECOM runs on the standard testproblem set

APL1P terminated by exceeding the bound on the samplesize.

The average elapsed time is 27.2 which is much lower than the elapsed time for QDECOM and DAPPROX; it would be even lower without the outlier coming from APL1P. Due to the stochastic nature of the algorithm underlying SDECOM, the result needs further statistical verification. One possibility is to build in statistically based stopping rules as proposed by Higle and Sen [59]. Otherwise, or if the algorithm terminates by a resource interrupt without fulfilling the statistical stopping criterion, the result needs statistical verification through subsequent sampling with x being fixed. The current version of SDE-COM does not contain the statistically based tests mentioned above. For this reason SDECOM will only be tested on some of the test problems and we will discuss the computational results only on the basis of the distance of the last incumbent to a solution returned by another solver. Remind that on this basis no conclusion can be made if this distance is relatively high.

Considering the computational results displayed in Table 6.9 we observe that SDECOM has found the same solution as DAPPROX for problems CEP1, SCRS8 and SCSD8.

TEST #3

The goal of this test is to consider the behavior of DAPPROX on a test problem with continuously distributed random variables. REFINE has been solved by DAPPROX with different levels of stopping accuracy. The results are shown in Table 6.10. Observe that both the overall number of diagonal LP's solved for computing the upper bound as those solved by QDECOM is very high for high (0.01%) accuracy. The computing time rapidly decreases for lower required accuracy levels. Notice that number of intervals of the subdivision becomes quite large for the 0.001% case. This implies that QDECOM, employed by DAPPROX for computing lower bounds, had to solve LP's with quite large numbers of diagonal blocks.

	LB on f	UB on f	t	ItCnt	Interv	ndiag LP	qdiag LP
0.01%	151.0414	151.0542	1036.5	56	8330	659064	753161
0.1%	151.0094	151.1530	179.3	36	1634	104762	140248
5%	149.6865	155.2241	6.8	13	77	2928	19138

Table 6.10: DAPPROX runs with REFINE; different stopping tolerances

	QDECOM			DAPPROX			
	f	t	ItCnt	UB - LB	t	ItCnt	Interv
CEP1X1	332232.418	9.8	21	0.0	28.4	27	81
CEP1X2	315090.674	9.7	20	0.0	24.9	25	89
CEP1X3	247135.815	10.4	22	$0.6 \cdot 10^{-10}$	26.6	24	90
CEP1X4	415873.180	10.4	23	0.0	40.7	37	85
CEP1X5	184960.705	10.5	22	$0.6 \cdot 10^{-10}$	24.3	22	89
CEP1X6	236742.483	10.0	20	$0.3 \cdot 10^{-10}$	26.5	23	89
CEP1X7	467020.392	10.8	24	0.0	32.9	30	84
CEP1X8	181937.910	11.3	23	$0.6 \cdot 10^{-10}$	26.9	23	86
CEP1X9	259287.110	9.9	20	$0.6 \cdot 10^{-10}$	30.7	27	86
CEP1X10	209727.213	13.1	28	$0.3 \cdot 10^{-10}$	34.0	25	81

Table 6.11: QDECOM and DAPPROX runs on the battery CEP1Xi

6.2.2   Variants of test problems from the literature

TEST #4

Our goal with this test is twofold: We wish to check the sensitivity of test problem CEP1 w.r. to the relatively high perturbation which leads to the test problem battery CEP1Xi, and on the other hand we would like to compare the performance of DAPPROX and QDECOM on these variants of CEP1. The stopping tolerance was 10^{-7} for DAPPROX and the objective values returned were equal up to 9 decimal digits. The computational results are shown in Table 6.11.

Notice that for each model both solvers have found the exact optimal point. Comparing solution times the runs show much lower computing times for QDECOM. When comparing the performance of the solvers with the base case model CEP1 (see Table 6.6), both of them show a very similar behavior for the original and for the perturbed model. The perturbation involves a very sensitive point of the model CEP1; observe the great variation in the optimal objective value.

TEST #5

This run is with the test battery PGP2X. The goals are anlogous of those for TEST #4 with the following additional feature: The stopping tolerance of DAPPROX will be set in these test runs, and in most of the test runs what follow to $\epsilon^* = 0.001$ (0.1%). Remind

	Δ	QDECOM			DAPPROX				
		f	t	ItCnt	LB on f	UB - LB	t	ItCnt	Interv
PGP2X1	6.0	451.151	43.9	48	451.000	0.364	11.2	15	35
PGP2X2	2.5	448.614	37.2	40	448.553	0.397	14.7	18	33
PGP2X3	1.5	446.891	46.0	54	446.816	0.322	10.4	14	29
PGP2X4	2.5	449.221	40.0	44	449.049	0.410	9.8	14	30
PGP2X5	0.0	444.384	45.4	53	444.280	0.430	11.8	15	32
PGP2X6	3.0	439.152	42.4	47	439.040	0.272	13.3	16	37
PGP2X7	1.5	436.972	38.7	43	436.914	0.323	13.5	18	33
PGP2X8	1.5	446.774	45.1	48	446.704	0.288	13.6	19	33
PGP2X9	2.5	451.101	37.1	40	450.970	0.341	11.0	15	37
PGP2X10	1.5	440.458	32.9	37	440.384	0.351	8.0	13	26

Table 6.12: QDECOM and DAPPROX runs on the battery PGP2Xi

	LB on f	UB on f	t	ItCnt	Interv	ndiag LP	qdiag LP
REFINED	150.549	151.671	36.6	24	178	27936	19138
REFINE1	154.052	154.193	156.1	35	1577	96984	113384

Table 6.13: DAPPROX runs with the models REFINED and REFINE1

that this means stopping according to the relative error, see Step 3 in Algorithm 3.1 on Page 70. This precision can be considered as high enough for most applications. The results of the run are shown in Table 6.12. As the perturbation now concerns the objective value, we also show the deviation of the solution obtained, in the column labeled by Δ, where $\Delta = \| x_B - x_P \|_\infty$ is the deviation of the solution x_B for the base case PGP2, and the solution x_P of the perturbed problem; both of them computed by QDECOM.

Comparing solution times now gives for DAPPROX the much better values due to the lower required accuracy. As the components of x_B lie between 1.5 and 5.5, the deviations in the last column indicate a strong response to the perturbation of q in the model PGP2.

TEST #6

The purpose of this test is to observe the effects of discretizing the distributions and of changing the type of distribution in REFINE. DAPPROX has been run on the model variants REFINED and REFINE1 with the stopping tolerance set to 0.1%. The results are shown in Table 6.13. Notice that the solution time for the discretized model RE-FINED is much lower than that for REFINE.

Observe that the relative error when comparing the lower bound of the discretized problem with that of the original one is 0.34%. The relative difference between the lower bound of the original problem and that of REFINE1 is 2% which may be significant as the objective of REFINE represents production costs in an aggregated manner, see Kall and Wallace [85].

	LB on f	UB on f	t	ItCnt	Interv	ndiag LP
CR1N1	23.0700	23.0781	624.5	18	100	322048
CR1N2	5.70885	6.01679	739.7	11	100	195584
CR1N3	6.52175	6.58226	552.7	12	100	186880
CR1N4	6.59578	6.60239	696.5	16	100	269312
CR1N5	2.96067	3.08204	639.7	10	100	187904
CR1N6	9.57522	9.57541	49.5	7	7	17920
CR1N7	15.6800	16.3225	883.7	12	100	216576
CR1N8	8.58576	8.58924	715.5	17	100	279552
CR1N9	1.88391	1.88414	29.5	5	5	10240
CR1N10	8.03476	8.03476	2.3	1	1	1024
CR1N11	6.24477	6.25586	912.1	15	100	285696
CR1N12	6.22184	6.28202	707.7	12	100	225280
CR1N13	12.8432	12.8432	2.5	1	1	1024
CR1N14	35.7484	35.7503	38.4	6	6	13824
CR1N15	14.5073	14.5073	2.7	1	1	1024
CR1N16	25.6635	25.7118	714.2	14	100	227840
CR1N17	0.109764	0.110890	792.6	13	100	282112
CR1N18	24.6966	24.7957	816.1	13	100	270848
CR1N19	16.8227	16.9496	611.9	11	100	181248
CR1N20	4.86273	4.88567	538.9	12	100	186368

Table 6.14: Results of a DAPPROX run with models CR1Ni

6.2.3   Randomly generated complete recourse problems

TEST #7

The goal of these runs was to consider the behavior of DAPPROX when applied to models with a relatively large number of continuously distributed random variables.

Test problems in the set CR1Ni were solved by DAPPROX. Remind that in this model the 10 independent components of the random RHS are normally distributed. On our 16MB machine this implied an upper bound equal to 100 for the subintervals.

The results are summarized in Table 6.14.

Some further characteristics of this test problem battery and the run are as follows (relative gap means $\frac{UB - LB}{1+ \mid LB \mid}$ with LB and UB being the lower and upper bounds, respectively).

- The relative gap at the expected value (starting) solution:

- Mean: 14%
- Minimum: 0% for CR1N10, CR1N13, CR1N15
- Maximum: 84% for CR1N9

The relative gap at the solution returned by DAPPROX:

- Mean: 0.9%
- Minimum: $\approx 10^{-14}$ for CR1N10, CR1N13, CR1N15
- Maximum: 5.4% for CR1N2

Observe the large overall number of diagonal LP's solved during the solution procedure. The stopping tolerance was 10^{-4} (0.01%). The solver terminated by fulfilling the termination criterion for CR1N6, CR1N9, CR1N10, CR1N13 an CR1N14; the rest of the runs was interrupted by exceeding the limit for the subdivisions.

TEST #8

The goal of the next group of runs was to consider the effect of a fine discretization of the continuous distributions on the objective values and on the performance of DAPPROX. The stopping tolerance of DAPPROX was 0.1% and the maximal number of subintervals in the subdivision was 70. These models have a large number (10^{10}) of joint realizations of the discretely distributed random variables.

Observe (Table 6.15) that the approximation yields very goood approximation to the optimal objective values of the original problem. Notice that now the optimality criterion has been fulfilled by more models as for the continuous case.

TEST #9

The purpose is to consider the behavior of SDECOM on the same problem set. The results are summarized in Table 6.16. The symbol Δ_{SD} in the header is the relative distance between the solutions obtained by DAPPROX and SDECOM, respectively, its definition can be found in Subsection 6.2.1. A zero value indicates that the two solutions are equal up to 8 decimal digits.

The table shows a remarkably good performance of SDECOM on this set of test problems.

TEST #10

The purpose of these runs is to investigate the behavior of DAPPROX on test problems in the battery CR2Xi. The 20 random entries are dependent for these models; both h and T are stochstaic. The number of random variables is 5. DAPPROX was started up with a stopping tolerance equal to 0.1% and the maximal number of subdivisions being 3000. Each of the runs terminated by exceeding this limit, so we omit this entry from the

	LB on f	UB on f	t	ItCnt	Interv	ndiag LP
CR1D1	23.0653	23.0770	192.1	9	20	74752
CR1D2	5.70450	5.89603	717.3	9	70	231424
CR1D3	6.52070	6.55325	885.8	15	70	389120
CR1D4	6.59215	6.55958	308.6	12	30	136192
CR1D5	2.96432	3.02378	653.1	10	70	226304
CR1D6	9.57471	9.58282	9.2	2	2	3072
CR1D7	15.6091	16.2304	928.6	10	70	282624
CR1D8	8.58362	8.59558	81.4	6	15	29696
CR1D9	1.87865	1.87865	26.8	4	4	10240
CR1D10	8.03471	8.03471	2.3	1	1	1024
CR1D11	6.24231	6.24708	741.4	12	70	254996
CR1D12	6.22083	6.25523	1045.7	13	70	388096
CR1D13	12.8428	12.8428	2.5	1	1	1024
CR1D14	35.7478	35.7478	2.6	1	1	1024
CR1D15	14.5073	14.5073	2.6	1	1	1024
CR1D16	25.6564	25.6825	561.4	12	55	217088
CR1D17	0.10916	0.110700	118.4	6	20	35840
CR1D18	24.6966	24.73356	1319.6	18	70	529408
CR1D19	16.8268	16.8863	670.9	11	70	256000
CR1D20	4.86167	4.87622	687.4	12	70	297984

Table 6.15: Results of a DAPPROX run with models CR1Di

table. The results are shown in Table 6.17.

Observe the large numbers of diagonal LP's which had to be solved by DAPPROX for obtaining the upper bound. Some further information of the run is summarized below:

- The relative gap at the starting solution:

    - Mean: 218%
    - Minimum: 160% for CR2X7
    - Maximum: 371% for CR2XX8

  The relative gap at the solution returned by DAPPROX:

    - Mean: 2.8%
    - Minimum: 1.6% for CR2X6
    - Maximum: 4.3% for CR2X17

TEST #11

The purpose of these runs is to consider the performance of DAPPROX and QDECOM on the battery CR2Di, which is a crude discretization of CR2Xi. For the computational

	Obj_est	t	Sas	Inc	Dv	Δ_{SD}
CR1D1	23.2180	20.1	215	61	4	$1.3 \cdot 10^{-2}$
CR1D2	5.74037	176.1	1000	307	12	$5.2 \cdot 10^{-2}$
CR1D3	6.48348	7.3	78	8	6	$1.0 \cdot 1^{-1}$
CR1D4	6.58896	95.7	1000	298	2	$3.2 \cdot 10^{-3}$
CR1D5	2.98520	23.6	256	48	8	0.0
CR1D6	9.63823	85.0	1000	268	2	0.0
CR1D7	15.8372	60.0	413	72	19	$2.8 \cdot 10^{-2}$
CR1D8	8.58134	3.1	50	7	2	0.0
CR1D9	1.92536	31.9	383	37	2	0.0
CR1D10	8.03471	2.8	50	1	1	0.0
CR1D11	6.22180	3.8	55	6	4	0.0
CR1D12	6.17631	7.6	100	17	4	0.0
CR1D13	12.8840	15.4	257	65	1	0.0
CR1D14	35.7478	2.9	50	1	1	$1.3 \cdot 10^{-8}$
CR1D15	14.5073	2.7	50	4	1	0.0
CR1D16	25.8693	112.4	1000	245	4	$1.6 \cdot 10^{-2}$
CR1D17	0.15463	8.0	139	3	1	0.0
CR1D18	24.2513	6.3	70	7	7	$2.0 \cdot 10^{-2}$
CR1D19	16.9154	155.5	1000	331	10	$6.0 \cdot 10^{-3}$
CR1D20	4.87983	3.6	50	3	3	0.0

Table 6.16: SDECOM runs on the battery CR1Di

results see Table 6.18.

Considering the optimal objective values, a large deviation can be observed in comparison with the optimal objective values of the original continuous models, see Table 6.17. This crude discretization leads to an inadequate approximation of the original problem.

From the point of view of solver behavior QDECOM performs much better on all of the problems in this test battery.

As a next step let us consider a finer discretization, leading to the family of test problems CR2Ei. For the results consult Table 6.19.

The approximation to the continuos problem became much better; let us consider solver performance.

Observe that the gap between the computing times of QDECOM and DAPPROX became smaller. There are even test problems now (CR2E2, CR2E5) for which the computing time of DAPPROX is the smaller.

The reason for the premature termination of QDECOM with problem CR2E20 is not yet clear; we made several trials with the cut tolerance varying between 10^{-12} and 10^{-5}.

	LB on f	UB on f	t	ItCnt	ndiag LP
CR2X1	170.076	173.772	982.6	30	285784
CR2X2	132.435	136.767	1293.2	28	309520
CR2X3	102.888	106.860	1728.7	27	258208
CR2X4	33.5160	34.5463	678.6	25	217856
CR2X5	142.569	146.853	1148.8	33	301248
CR2X6	87.0642	88.4707	793.5	27	250992
CR2X7	65.5961	67.639	860.5	29	259504
CR2X8	41.1940	42.8576	1010.2	28	299152
CR2X9	141.870	144.7826	1008.4	32	323296
CR2X10	153.711	156.3948	673.0	25	218496
CR2X11	176.767	182.1705	1146.2	28	250272
CR2X12	38.6322	39.7078	855.0	24	216016
CR2X13	233.054	237.823	907.3	29	272800
CR2X14	63.9720	65.1552	1441.6	32	374416
CR2X15	131.222	136.747	1055.9	24	237840
CR2X16	121.160	125.478	1347.0	31	318672
CR2X17	84.7420	88.3202	1453.2	30	316656
CR2X18	125.864	127.936	1141.8	25	268576
CR2X19	80.8788	82.9155	1141.9	28	308752
CR2X20	167.506	170.275	1015.9	27	240352

Table 6.17: Results of a DAPPROX run with models CR2Xi

Let us make the discretization even finer, i.e. let us consider the test battery CR2Fi. The results are shown in Table 6.20.

We observe that the objective approximation did not improve too much. The relative performance of the two solvers however changed a lot. Their behavior from the efficiency point of view became very similar, the advantages of QDECOM concerning computing speed diminished.

6.2.4   Randomly generated simple recourse problems with only the RHS being stochastic

__TEST #12__

Goal of the test is to demonstrate that for simple recourse problems with only the RHS being stochastic very large problems can be solved efficiently.

SRAPPROX has been first run on the test-problem family SR1Ni. The stopping tolerance for the relative difference in the bounds was 10^{-6}. The results are summarized in

	QDECOM			DAPPROX				
	f	t	ItCnt	LB on f	UB - LB	t	ItCnt	Interv
CR2D1	160.574	13.5	20	160.517	0.081	70.2	24	88
CR2D2	125.640	31.4	54	125.543	0.11	90.8	24	118
CR2D3	98.1424	23.8	34	98.0750	0.087	132.4	22	113
CR2D4	32.3183	8.6	14	32.2899	0.0332	71.7	22	97
CR2D5	136.201	11.0	37	136.110	0.124	68.0	20	86
CR2D6	83.198	4.6	5	83.1305	0.076	43.3	18	66
CR2D7	63.0246	16.0	21	62.9790	0.058	90.0	24	99
CR2D8	38.9718	6.4	8	38.9438	0.032	82.9	22	109
CR2D9	135.647	4.1	7	135.582	0.091	33.4	17	59
CR2D10	145.758	7.6	10	145.678	0.105	78.9	25	96
CR2D11	168.550	10.0	12	168.421	0.158	120.1	24	125
CR2D12	36.8673	6.0	7	36.8355	0.035	86.2	22	99
CR2D13	219.961	7.6	8	219.843	0.152	66.1	23	80
CR2D14	60.7132	21.5	34	60.6729	0.046	47.7	19	75
CR2D15	125.893	13.2	14	125.797	0.125	47.0	16	69
CR2D16	115.329	18.0	23	115.248	0.111	99.5	25	109
CR2D17	81.1940	16.0	26	81.1502	0.063	144.0	30	112
CR2D18	120.001	9.8	12	119.919	0.106	49.4	19	67
CR2D19	77.1012	10.4	15	77.0368	0.073	102.3	25	107
CR2D20	158.915	13.9	14	158.815	0.133	74.1	22	86

Table 6.18: QDECOM and DAPPROX runs on the battery CR2Di

	QDECOM			DAPPROX				
	f	t	ItCnt	LB on f	UB - LB	t	ItCnt	Interv
CR2E1	167.322	309.0	35	167.197	0.160	373.3	32	478
CR2E2	130.496	1140.6	183	130.408	0.110	665.9	37	671
CR2E3	101.720	455.8	55	101.657	0.074	1442.4	39	844
CR2E4	33.2263	170.5	13	33.1958	0.034	579.5	34	644
CR2E5	140.528	667.4	92	140.444	0.105	510.4	37	442
CR2E6	85.7407	79.6	6	85.6791	0.076	313.1	29	362
CR2E7	64.8610	168.1	25	64.8055	0.063	589.6	39	505
CR2E8	40.5858	245.5	22	40.5581	0.033	784.5	40	646
CR2E9	140.179	92.1	15	140.071	0.139	378.7	35	363
CR2E10	152.399	271.7	33	152.304	0.113	423.9	32	514
CR2E11	174.064	339.8	38	173.949	0.129	970.8	41	811
CR2E12	38.1417	152.9	21	38.1082	0.038	572.1	31	600
CR2E13	228.075	95.9	6	227.892	0.227	416.9	33	501
CR2E14	63.2448	359.8	50	63.2005	0.054	550.1	39	422
CR2E15	129.927	168.5	15	129.839	0.119	418.2	29	505
CR2E16	119.896	325.7	35	119.802	0.111	560.4	35	482
CR2E17	83.7716	368.6	49	83.7029	0.078	781.5	38	598
CR2E18	124.081	195.9	21	123.992	0.112	309.8	29	354
CR2E19	79.6613	208.3	24	79.6103	0.064	849.6	39	667
CR2E20	258.049	373.8	36	165.357	0.153	304.2	33	271

Table 6.19: QDECOM and DAPPROX runs on the battery CR2Ei

	QDECOM			DAPPROX				
	f	t	ItCnt	LB on f	UB - LB	t	ItCnt	Interv
CR2F1	167.501	1250.9	33	167.414	0.146	687.5	33	848
CR2F2	130.986	4782.4	254	130.918	0.104	1619.5	40	1543
CR2F3	102.386	1998.6	70	102.316	0.094	2376.7	40	1644
CR2F4	33.2305	928.5	14	33.2060	0.034	1217.7	38	1311
CR2F5	141.693	2795.6	99	141.618	0.112	1125.9	45	942
CR2F6	86.0330	383.8	6	85.9897	0.072	437.0	30	628
CR2F7	65.0173	564.1	24	64.9708	0.064	909.2	37	940
CR2F8	41.0291	1359.9	31	41.0081	0.028	1510.3	39	1399
CR2F9	139.5473	331.3	14	139.4748	0.125	637.7	36	795
CR2F10	151.189	1232.5	39	151.0910	0.148	680.4	34	913

Table 6.20: QDECOM and DAPPROX runs on the battery CR2Fi

	Overall t	Phase-I t	Phase-II t	ItCnt	f
SR1N1	101.88	84.31	17.57	11	275.919
SR1N2	105.57	80.90	24.67	16	103.372
SR1N3	87.23	76.51	10.72	11	367.819
SR1N4	101.83	89.20	12.63	13	160.198
SR1N5	116.6	101.39	15.21	11	365.579
SR1N6	102.49	91.89	10.6	11	397.430
SR1N7	106.33	96.39	9.94	11	278.995
SR1N8	92.44	79.59	12.85	11	366.953
SR1N9	102.33	91.23	11.1	11	282.112
SR1N10	94.96	80.68	14.28	13	165.586
SR1N11	95.19	82.33	12.86	11	217.385
SR1N12	122.38	108.92	13.46	11	306.381
SR1N13	110.67	100.51	10.16	11	180.316
SR1N14	98.32	84.37	13.95	13	191.566
SR1N15	99.91	87.77	12.14	10	346.946
SR1N16	109.35	97.66	11.68	12	227.532
SR1N17	117.93	98.75	19.18	15	181.211
SR1N18	98.82	82.11	16.71	11	339.628
SR1N19	108.26	93.65	14.61	11	266.271
SR1N20	110.51	96.67	13.84	11	267.377

Table 6.21: Results of a SRAPPROX run with models SR1Ni

	Overall t	Phase-I t	Phase-II t	ItCnt	f
SR1D1	94.31	84.09	10.22	6	274.72935
SR1D2	93.54	80.80	12.74	6	102.661504
SR1D3	81.57	76.45	5.12	6	366.699603
SR1D4	94.75	89.20	5.55	5	159.383103
SR1D5	118.15	110.02	8.13	6	364.412119

Table 6.22: Results of a SRAPPROX runs with models SR1Di

Table 6.21.

In the next run SRAPPROX was used to solve the test-problems family SR1Di. These are problems with finitely distributed random variables. The stopping was the same as before, but in this case at stopping the accuracy obtained was much better (10^{-12}) due to the discrete distributions. Notice that the number of joint realizations is 10^{90} for the models in SR1Di. The results are shown in Table 6.22.

Observe that the Phase II computing times are quite low in both Tables 6.21 and 6.22. The Phase I time is just for solving the starting expected value problem. Afterwards the hot start described in Section 3.1 is used which results in a quite speedy solver.

6.3   Test runs for jointly chance constrained problems

The default parameter settings for the solvers are the following:

- *PCSPIOR:* The maximal number of cuts is 200; the sample size for Monte Carlo integration is 1000 and the stopping tolerance is 10^{-6}.

- *PROBALL:* The maximal number of cuts is 200; a cut is considered inactive if it is inactive in 3 subsequent iterations; slack cuts are dropped at a period of 10 iterations. The sample size for Monte Carlo integration is 1000 and the stopping tolerance is 10^{-6}.

- *PROCON:* The sample size for Monte Carlo integration is 1000 and the stopping tolerance is 10^{-6}.

All of the three solvers are recently under development, which implies that conclusions concerning the undelying algorithms can only be made in taking this fact into consideration. For all of the three algorithms the first phase, i.e. finding a starting Slater point is implemented by using the heuristic algorithm 3.10 on Page 84. For implementing an exact Phase I algorithm further testing is needed to select the best underlying algorithm.

In the tables of this section we will use the following abbreviations: "P_{LB}" and "P_{UB}" denote lower and upper bounds on the probability corresponding to the solution (i.e. on the value of the distribution function $F(x^*)$ at the solution x^*), respectively; "FCnt" stands for the number of evaluations of the probability distribution function (see the jointly chance constrained model formulation (2.23) on Page 60).

6.3.1   Test problems from the literature

TEST #13

The goal of these runs is to consider the performance of the solvers on the standard test problems selected.

The three solvers PCSPIOR, PROBALL and PROCON were run on the test problems STABIL and WATRES, both with prescribed probability level equal to 0.9. The computational results are shown in Tables 6.23 and 6.24.

The solvers PCSPIOR, PROBALL and PROCON were also run on STABIL with prescribed probability level equal to 0.95. The computational results are presented in Tables 6.25 and 6.26.

	PCSPIOR			PROBALL			PROCON		
	f	P_{LB}	P_{UB}	f	P_{LB}	P_{UB}	f	P_{LB}	P_{UB}
STABIL	4370.29	.898	.900	4370.24	.901	.903	4370.28	.897	.900
WATRES	394.886	.898	.901	394.886	.956	.957	394.886	.976	.976

Table 6.23: Standard test problems; $\alpha = 0.9$

	PCSPIOR			PROBALL			PROCON		
	t	ItCnt	FCnt	t	ItCnt	FCnt	t	ItCnt	FCnt
STABIL	12.9	14	155	4.7	13	28	6.2	50	72
WATRES	3.0	6	48	1.0	2	11	0.4	4	0

Table 6.24: Standard test problems; $\alpha = 0.9$

	PCSPIOR			PROBALL			PROCON		
	f	P_{LB}	P_{UB}	f	P_{LB}	P_{UB}	f	P_{LB}	P_{UB}
STABIL	4369.18	.949	.950	4368.98	.955	.956	4369.13	.949	.950

Table 6.25: Standard test problems; $\alpha = 0.95$

	PCSPIOR			PROBALL			PROCON		
	t	ItCnt	FCnt	t	ItCnt	FCnt	t	ItCnt	FCnt
STABIL	12.6	13	141	2.7	5	19	5.2	49	52

Table 6.26: Standard test problems; $\alpha = 0.95$

	PCSPIOR			PROBALL			PROCON		
	f	P_{LB}	P_{UB}	f	P_{LB}	P_{UB}	f	P_{LB}	P_{UB}
STABIL1	4314.93	.901	.904	4314.94	.901	.903	4315.02	.897	.900
STABIL2	4345.05	.903	.905	4345.10	.901	.903	4342.60	.898	.900
STABIL3	4358.25	.898	.500	4358.06	.905	.907	4358.17	.895	.900
STABIL4	4543.26	.898	.500	4543.16	.903	.905	4543.03	.895	.900
STABIL5	4409.03	.906	.909	4409.10	.903	.905	4409.16	.896	.900
STABIL6	4514.98	.902	.904	4514.88	.908	.910	4515.05	.897	.900
STABIL7	3972.41	.000	.006	3871.78	.900	.903	3971.81	.896	.900
STABIL8	4300.47	.900	.902	4300.47	.900	.902	4297.26	.899	.900
STABIL9	4289.28	.898	.900	4289.23	.900	.903	4389.05	.895	.900
STABIL10	4663.20	.898	.900	4662.86	.916	.918	4663.19	.897	.900

Table 6.27: Perturbed STABIL model

	PCSPIOR			PROBALL			PROCON		
	t	ItCnt	FCnt	t	ItCnt	FCnt	t	ItCnt	FCnt
STABIL1	9.3	11	117	6.7	15	54	9.7	141	54
STABIL2	9.2	10	107	6.6	16	53	2.8	34	14
STABIL3	13.6	14	160	4.7	13	28	12.0	182	66
STABIL4	12.9	13	148	4.6	14	28	5.5	43	62
STABIL5	9.9	11	119	8.4	38	38	7.3	64	83
STABIL6	8.4	9	97	3.0	7	19	5.7	37	82
STABIL7	7.5	8	83	9.2	18	99	7.7	57	102
STABIL8	12.9	14	166	6.9	15	55	12.4	171	81
STABIL9	14.8	15	174	6.5	15	54	5.2	37	64
STABIL10	14.7	15	171	2.9	7	19	9.1	64	126

Table 6.28: Perturbed STABIL model

6.3.2   Variants of test problems from the literature

TEST #14

The purpose of this test is to consider solver behavior on, and optimal objective values of STABIL for variants emerging from the perturbation described in subsection 6.1.5. The probability level is $\alpha = 0.9$. The results are shown in Tables 6.27 and 6.28.

The solvers show a similar behavior as for the original unperturbed model; the optimal objective has large deviations from the base case value. Observe that for testproblem STABIL7 PCSPIOR terminated with an infeasible solution. This is due to a heuristic termination criterion based on small changes of the objective in a certain number of subsequent iterations. This does not work for STABIL7 which apparently is rather insensitive to the stochastic constraint. This heuristic stopping rule must be improved in PCSPIOR.

		PCSPIOR			PROCON		
	f^*	f	P_{LB}	P_{UB}	f	P_{LB}	P_{UB}
J1CC1	3263.82	3263.79	.900	.901	3263.91	.901	.901
J1CC2	876.265	876.242	.900	.900	876.689	.906	.906
J1CC3	3785.96	3785.89	.900	.900	3806.34	.900	.900
J1CC4	2966.01	2965.95	.900	.900	2967.05	.904	.904
J1CC5	1002.37	1002.34	.899	.900	1002.38	.898	.900
J1CC6	1815.27	1815.22	.900	.900	1832.74	.900	.900
J1CC7	3221.64	3221.46	.900	.900	3284.77	.900	.900
J1CC8	2971.77	2971.77	.896	.900	3812.53	.917	.917
J1CC9	2821.74	2821.81	.900	.900	2821.54	.903	.903
J1CC10	2877.16	2876.97	.804	.807	2886.48	.903	.903
J1CC11	4647.53	4647.41	.900	.900	4647.53	.900	.900
J1CC12	4413.87	4414.07	.900	.900	4413.93	.900	.900
J1CC13	1846.29	1846.25	.899	.900	1846.39	.899	.900
J1CC14	4859.97	4859.87	.900	.900	4861.04	.901	.901
J1CC15	4070.24	4070.26	.900	.900	4069.51	.907	.907
J1CC16	2820.88	2820.90	.898	.900	2811.15	.895	.900
J1CC17	182.904	182.872	.898	.900	183.022	.898	.900
J1CC18	2073.15	2073.16	.899	.900	2072.87	.902	.903
J1CC19	608.470	608.500	.900	.900	608.292	.900	.900
J1CC20	1635.325	1635.04	.899	.900	1635.48	.899	.900

Table 6.29: PCSPIOR and PROCON runs on models J1CCi

6.3.3 Randomly generated test problems

TEST #15

The goal of this test is to consider solver behavior on the randomly generated test battery J1CCi. As for these models the optimal objective value is known, another purpose is to check the accuracy of the objective value returned by the solver.

The probability level is $\alpha = 0.9$. For this set of test problems PCSPIOR and PROCON have been run. The computational results are summarized in Tables 6.29 and 6.30. The first column with the header f^* contains the optimal objective values as returned by GENSLP.

Observe that PCSPIOR returns objective values which closely approximate the exact value for each of the test problems. The implementation of PROCON does not strive to achieve very accurate results. The same comment, made for two stage problems concerning required accuracy, also applies for chance constrained problems: A 0.1% relative accuracy should be sufficient for most of the applications.

As mentioned earlier, the development of the three solvers PCSPIOR, PROBALL and

	PCSPIOR			PROCON		
	t	ItCnt	FCnt	t	ItCnt	FCnt
J1CC1	14.3	15	76	15.9	240	67
J1CC2	3.9	4	0	4.7	1	0
J1CC3	5.1	7	14	7.0	98	0
J1CC4	3.7	5	1	3.0	40	0
J1CC5	26.1	19	146	31.1	298	18
J1CC6	16.4	18	49	16.0	260	0
J1CC7	13.2	15	63	13.8	263	0
J1CC8	58.9	42	380	24.8	244	100
J1CC9	9.0	11	32	8.0	133	0
J1CC10	7.4	8	6	7.6	140	0
J1CC11	14.6	15	73	14.8	166	12
J1CC12	8.0	11	6	12.5	206	0
J1CC13	16.8	17	111	10.3	150	29
J1CC14	5.5	6	22	3.5	46	0
J1CC15	7.0	7	26	5.6	89	0
J1CC16	34.6	29	235	16.6	100	99
J1CC17	22.2	18	124	8.9	149	12
J1CC18	22.9	19	146	13.2	233	1
J1CC19	11.0	15	44	10.2	188	90
J1CC20	33.0	26	216	6.8	97	103

Table 6.30: PCSPIOR and PROCON runs on models J1CCi

PROCON is not yet completed. Our tests concerning jointly chance constrained problems aim in the first line the further development of the solvers.

The accuracy of the results computed by PROCON is not yet satisfactory; consider e.g. J1CC8 in Table 6.29. Let us point out that the solver PROCON, which participates in the comparisons of this study, is not the same as the solver described in Mayer [108]. It is a completely different implementation of the algorithm presented in [108].

6.4   Conclusions

6.4.1   Two stage problems

Discrete distribution, complete recourse

The effect of the number of realizations

The number of the joint realizations of the random variables clearly sets a limit to the applicability of algorithms based on solving the algebraic equivalent LP. Examples are problems in the test batteries CR1Di with 10^{10} realizations and the models belonging to the family SR1Di with 10^{90} realizations. None of these models can be solved by the algebraic equivalent LP approach. The reason is twofold: On the one hand available storage capacity of present day machines does not allow for setting up the deterministic equivalent LP for such big models. While in the case of CR1Di computer technology development may lead to machines which have the necessary storage capacity, the author has his doubts whether there it will ever be possible to apply the algebraic equivalent LP approach to models in the family SR1Di. Lack of storage capacity is only one of the problems. If we assume e.g. that the models in CR1Di can be set up, the problem still remains how to solve them. These LP's would involve hundreds of billions of rows and columns which is far beyond present day experience with LP's.

TEST #10 in Subsection 6.2.3 shows that even the very efficient regularized decomposition approach has a decreasing relative efficiency with increasing number of joint realizations, when compared to a discrete approximation method. This may be a general trend; its justification needs however further experimentation with larger samplesizes and statistically based reasoning.

The effect of the number of random variables

The number of random variable sets a limit to approximation methods based on solving second stage LP's on the vertices of a rectangular subdivision. The exponential relation between the number of vertices and the number of random variables imposes a limit on this approach lying somewhere around 10 random variables. Observe the large amount of diagonal LP's needed for solving models in the test batteries CR1Ni and CR1Di (Tests #7 and #8).

The effect of the required accuracy

For stochastic programming problems a 0.1% relative accuracy is in most cases considered to be high. For this reason it was fair in our tests to run DAPPROX with this required accuracy in the comparative contexts. If a higher precision is needed then either a solver based on the algebraic equivalent LP, or a solver based on an approximation scheme is to be used. In the letter case repeated startup based on hotstart may be needed. Notice that the computing time rapidly increases for DAPPROX in all cases considered in this study, when the required precision increases.

The general fixed recourse case

A great advantage of the algebraic equivalent approach is that it also works in the case of incomplete recourse. In the relatively complete recourse case DAPPROX and SDECOM can also be applied, see the tests #1 and 2.

Continuous distribution, complete recourse

One possible approach is to use a deterministic or stochastic approximation method, e.g. solvers DAPPROX and SDECOM. Another approach consists of discretizing the distributions and solving afterwards the models now having discretely distributed random variables. The second approach may result in problems having a large amount of joint realizations, see Test #10 for the accuracy achieved through this type of approximations.

The simple recourse case

For simple recourse problems, solvers utilizing the separable structure are much more efficient then general solvers. This obvious effect has not been shown in the study in the comparative context. Notice however that SRAPPROX solved huge models (SR1Ni and SR1Di) quite efficiently, which are out of the range of DAPPROX or of QDECOM in the case of SR1Di. SDECOM is applicable to these models but the author doubts that it can be as efficient as an approach based on the separability property.

6.4.2   Jointly chance constrained problems

Due to the fact that the solvers participating in this study are not yet completed, we confine ourselves to identifying some general patterns.

PCSPIOR is based on an outer approximation method and turned out to be quite robust in the tests. The weakness of this approach is that requiring lower precision (most probably needed for large scale problems) results in infeasible solutions at stopping.

The central cutting plane method has the following obvious advantage: In iterations where an objective cut is carried out, the gradient of the stochastic constraint need not to be computed. A further advantage is that for lower required accuracy a feasible solution is returned.

The reduced gradient method has the advantage of throughout working with feasible points, so it is also well suited to stopping with a lower required accuracy. The weakness of the approach is that it needs quite accurately computed gradients for the stochastic constraint if this constraint is active.

Bibliography

[1] J. Abadie and J. Carpentier. Généralisation de la méthode du gradient réduit de Wolfe au cas des contraintes non-linéaires. In D. B. Hertz and J. Melese, editors, *Proceedings of an IFORS Conference*, pages 1041–1053. John Wiley & Sons, 1966.

[2] O. Bahn, O. du Merle, J. L. Goffin, and J. P. Vial. A cutting plane method from analytic centers for stochastic programming. Cahiers de recherche 1993.5, Université de Genéve, 1993.

[3] H. Bauer. *Wahrscheinlichkeitstheorie*. Walter de Gruyter, 1991.

[4] M. S. Bazaraa and C. M. Shetty. *Nonlinear programming. Theory and algorithms*. John Wiley & Sons, 1979.

[5] J. F. Benders. Partitioning procedures for solving mixed variables programming problems. *Numerische Mathematik*, 4:238–252, 1962.

[6] A. Bharadwaj, J. Choobineh, Lo A., and B. Shetty. Model management systems: A survey. *Annals of Operations Research*, 38:17–67, 1992.

[7] J. R. Birge. The value of stochastic solution in stochastic linear programs with fixed recourse. *Mathematical Programming*, 24:314–325, 1982.

[8] J. R. Birge. Decomposition and partitioning methods for multistage stochastic linear programs. *Operations Research*, 33:989–1007, 1985.

[9] J. R. Birge, M. A. H. Dempster, H. Gassmann, E. Gunn, A. J. King, and S. W. Wallace. A standard input format for multiperiod stochastic linear programs. Working Paper WP-87-118, IIASA, 1987.

[10] J. R. Birge and D. Holmes. Efficient solution of two stage stochastic linear programs using interior point methods. *Computational Optimization and Applications*, 1:245–276, 1992.

[11] J. R. Birge and F. V. Louveaux. A multicut algorithm for two stage linear programs. *European Journal of Operational Research*, 34:384–392, 1988.

[12] J. R. Birge and R. J-R. Wets. Designing approximation schemes for stochastic optimization problems, in particular for stochastic programs with recourse. *Mathematical Programming Study*, 27:54–102, 1986.

[13] J.J. Bisschop and A. Meeraus. On the development of a general algebraic modeling system in a strategic planning environment. *Mathematical Programming Study*, 20:1–29, 1982.

[14] C. Borell. Convex set-functions in *d*-space. *Period. Math. Hungar.*, 6:111–136, 1975.

[15] E. Boros and A. Prékopa. Closed-form two-sided bounds for probabilities that at least r and exactly r out of n events occur. *Mathematics of Operations Research*, 14:317–342, 1989.

[16] H. J. Brascamp and E. H. Lieb. On extensions of the Brunn-Minkowski and Prékopa-Leindler theorems, including inequalities for log concave functions, and with an application to the diffusion equation. *Journal of Functional Analysis*, 22:366–389, 1976.

[17] A. Brooke, D. Kendrick, and A. Meeraus. *GAMS. A User's Guide, Release 2.25.* Boyd and Fraser/The Scientific Press, Danvers, MA, 1992.

[18] H. P. Crowder, R. S. Dembo, and J. M. Mulvey. Reporting computational experiments in mathematical programming. *Mathematical Programming*, 15:316–329, 1978.

[19] G. B. Dantzig. *Linear programming and extensions.* Princeton University Press, 1963.

[20] G. B. Dantzig and P. W. Glynn. Parallel processors for planning under uncertainty. *Annals of Operations Research*, 22:1–22, 1990.

[21] I. Deák. Three digit accurate multiple normal probabilities. *Numerische Mathematik*, 35:369–380, 1980.

[22] I. Deák. Multidimensional integration and stochastic programming. In Y. Ermoliev and R.J-B. Wets, editors, *Numerical Techniques for Stochastic Optimization*, pages 187–200. Springer Verlag, 1988.

[23] I. Deák. *Random number generators and simulation.* Akadémiai Kiadó, Budapest, 1990.

[24] R. S. Dembo and J. G. Klincewicz. Dealing with degeneracy in reduced gradient algorithms. *Mathematical Programming*, 31:357–363, 1985.

[25] R. S. Dembo and J. M. Mulvey. On the analysis and comparison of mathematical programming algorithms and software. In W. W. White, editor, *Computers and mathematical programming*, pages 106–116. National Bureau of Standards, Washington, D.C., 1978.

[26] D. R. Dolk. Model management systems for operations research: A prospectus. In G. Mitra, editor, *Mathematical Methods for Decision Support*, pages 347–373. Springer Verlag, 1988.

[27] J. Dupačová. Minimax stochastic programs with nonconvex nonseparable penalty functions. In A. Prékopa, editor, *Progress in Operations Research*, pages 303–316. North-Holland Publ. Co., 1974.

[28] J. Dupačová. Minimax stochastic programs with nonseparable penalties. In K. Iracki, K. Malanowski, and S. Walukiewicz, editors, *Optimization Techniques, Part I*, pages 157–163. Springer Verlag, 1980.

[29] J. Dupačová. The minimax approach to stochastic programming and an illustrative application. *Stochastics*, 20:73–88, 1987.

[30] J. Dupačová, A. Gaivoronski, Z. Kos, and T. Szántai. Stochastic programming in water management: A case study and a comparison of solution techniques. *European Journal of Operational Research*, 52:28–44, 1991.

[31] J. Elzinga and T. G. Moore. A central cutting plane method for the convex programming problem. *Mathematical Programming*, 8:134–145, 1975.

[32] Y. Ermoliev. Stochastic quasigradient methods and their application to systems optimization. *Stochastics*, 9:1–36, 1983.

[33] Y. Ermoliev. Stochastic quasigradient methods. In Y. Ermoliev and R.J-B. Wets, editors, *Numerical Techniques for Stochastic Optimization*, pages 143–185. Springer Verlag, 1988.

[34] A. V. Fiacco and G. P. McCormick. *Nonlinear programming: Sequential unconstrained minimization techniques*. John Wiley & Sons, 1968.

[35] R. Fikes and T. Kehler. The role of frame-based representation in reasoning. *Communications of the ACM*, 28:904–920, 1985.

[36] L. D. Fosdick, editor. *Performance evaluation of numerical software*. North-Holland Publ. Co., 1979.

[37] K. Frauendorfer. Solving SLP recourse problems with arbitrary multivariate distributions - the dependent case. *Mathematics of Operations Research*, 13:377–394, 1988.

[38] K. Frauendorfer and P. Kall. A solution method for SLP recourse problems with arbitrary multivariate distributions — the independent case. *Problems of Control and Information Theory*, 17:177–205, 1988.

[39] K. Frauendorfer. *Stochastic two-stage programming*. Springer Verlag, 1992.

[40] K. R. Frisch. The logarithmic potential method for solving linear programming problems. Memorandum, Institute of Economics, Oslo, 1955.

[41] K. R. Frisch. The logarithmic potential method of convex programming. Memorandum, Institute of Economics, Oslo, 1955.

[42] O. E. Flippo and H. G. Rinooy Kan. Decomposition in general mathematical programming. *Mathematical Programming*, 60:361–382, 1993.

[43] A. Gaivoronski. Stochastic quasigradient methods and their implementation. In Y. Ermoliev and R.J-B. Wets, editors, *Numerical Techniques for Stochastic Optimization*, pages 313–351. Springer Verlag, 1988.

[44] A. Gaivoronski. Interactive program SQG-PC for solving stochastic programming problems on IBM/XT/AT compatibles —User Guide—. Working Paper WP-88-11, IIASA, 1988.

[45] H. I. Gassmann and A. M. Ireland. Scenario formulation in an algebraic modelling language. Working Paper WP-92-7, School of Business Administration, Dalhousie University, Halifax, 1992.

[46] H. I. Gassmann and A. M. Ireland. On the formulation of stochastic linear programs using algebraic modeling languages. *Annals of Operations Research*, 64:83–112, 1996.

[47] A. M. Geoffrion. Duality in nonlinear programming: A simplified applications-oriented approach. *SIAM Review*, 13:1–37, 1970.

[48] A.M. Geoffrion. Primal resource-directive approaches for optimizing nonlinear decomposable systems. *Operations Research*, 18:375–403, 1970.

[49] A.M. Geoffrion. Generalized Benders-decomposition. *Journal of Optimization Theory and Applications*, 10:237–260, 1972.

[50] J. L. Goffin and J. P. Vial. Cutting planes and column generation techniques with the projective algorithm. *Journal of Optimization Theory and Applications*, 65:409–429, 1988.

[51] I. Graham. *Object oriented methods*. Addison-Wesley Publ. Co., 1994.

[52] H. J. Greenberg. *A computer-assisted analysis system for mathematical programming models and solutions: A user's guide to ANALYZE*. Kluwer Academic Publishers, 1993.

[53] C. Grossmann and H. Kleinmichel. *Verfahren der nichtlinearen Optimierung*. Teubner, Leipzig, 1976.

[54] D. den Hertog. *Interior-point approach to linear, quadratic and convex programming: algorithms and complexity*. Kluwer Academic Publishers, 1994.

[55] D. den Hertog, J. Kaliski, C. Roos, and T. Terlaky. A note on central cutting plane methods for convex programming. Report 93-44, Delft University of Technology, Faculty of Technical Mathematics and Informatics, 1993.

[56] D. den Hertog, J. Kaliski, C. Roos, and T. Terlaky. A logarithmic barrier cutting plane method for convex programming. *Annals of Operations Research*, 58:69–98, 1995.

[57] J. L. Higle and S. Sen. Stochastic decomposition: An algorithm for two-stage linear programs with recourse. *Mathematics of Operations Research*, 16:650–669, 1991.

[58] J. L. Higle and S. Sen. Guidelines for a computer implementation of stochastic decomposition algorithms. Technical report, Systems and Industrial Engineering Department, University of Arizona, Tucson, February 1991.

[59] J. L. Higle and S. Sen. Statistical verification of optimality conditions for stochastic programs with recourse. *Annals of Operations Research*, 30:215–240, 1991.

[60] J. L. Higle and S. Sen. Finite master programs in regularized stochastic decomposition. *Mathematical Programming*, 67:143–168, 1994.

[61] J. L. Higle, W. W. Lowe, and R. Odio. Conditional stochastic decomposition: An algorithmic interface for optimization and simulation. *Operations Research*, 42:311–322, 1994.

[62] J. L. Higle and S. Sen. *Stochastic decomposition. A statistical method for large scale stochastic linear programming*. Kluwer Academic Publishers, 1996.

[63] J. K. Ho and E. Loute. A set of staircase linear programming test problems. *Mathematical Programming*, 20:245–250, 1981.

[64] W. W. Hogan. Applications of a general convergence theory for outer approximation algorithms. *Mathematical Programming*, 5:151–168, 1973.

[65] W. W. Hogan. Point-to-set maps in mathematical programming. *SIAM Review*, 15:591–603, 1973.

[66] D. Holmes. A collection of stochastic programming problems. Technical report 94-11, Department of Industrial and Operations Engineering, The University of Michigan, Ann Arbor, April 1994.

[67] G. Infanger. Monte carlo (importance) sampling within a benders decomposition algorithm for stochastic linear programs. *Annals of Operations Research*, 39:69–95, 1992.

[68] G. Infanger. *Planning under uncertainty: Solving large-scale stochastic linear programs*. Boyd & Fraser Publ. Co., Danvers, MA, 1994.

[69] F. Jarre. Interior-point methods via self-concordance or relative lipschitz-condition. Habilitationsschrift, Fakultät für Mathematik, Universität Würzburg, 1994.

[70] P. Kall. Approximations to stochastic programs with complete fixed recourse. *Numerische Mathematik*, 22:333–339, 1974.

[71] P. Kall. *Stochastic linear programming*. Springer Verlag, 1976.

[72] P. Kall. *Mathematische Methoden des Operations Research*. Teubner, 1976.

[73] P. Kall. Computational methods for solving two-stage stochastic linear programming problems. *Zeitschrift für angewandte Mathematik und Physik*, 30:261–271, 1979.

[74] P. Kall. Stochastic programs with recourse: an upper bound and the related moment problem. *Zeitschrift für Operations Research*, 31:A119–A141, 1987.

[75] P. Kall. On approximation and stability in stochastic programming. In J. Guddat, H. Th. Jongen, B. Kummer, and F. Nožička, editors, *Parametric Optimization and Related Topics*, pages 387–407. Akademie-Verlag, 1987.

[76] P. Kall. Stochastic programming with recourse: Upper bounds and moment problems – a review. In J. Guddat, B. Bank, H. Hollatz, P. Kall, D. Klatte, B. Kummer, K. Lommatzsch, K. Tammer, M. Vlach, and K. Zimmermann, editors, *Advances in Mathematical Optimization (Dedicated to Prof. Dr.Dr.hc. F. Nožička)*, pages 86–103. Akademie-Verlag, Berlin, 1988.

[77] P. Kall. An upper bound for SLP using first and total second moments. *Annals of Operations Research*, 30:267–276, 1991.

[78] P. Kall. A review on approximations in stochastic programming. In V. P. Bulatow, editor, *Optimization: Models, methods and solutions; Proceedings of the 1989 Baikal Conference on Global Optimization*, pages 125–133. Nauka, Nowosibirsk, 1992. in Russian.

[79] P. Kall and J. Mayer. SLP-IOR: A model management system for stochastic linear programming — system design —. In A.J.M. Beulens and H.-J. Sebastian, editors, *Optimization-Based Computer-Aided Modelling and Design*, pages 139–157. Springer Verlag, 1992.

[80] P. Kall and J. Mayer. A model management system for stochastic linear programming. In P. Kall, editor, *System Modelling and Optimization*, pages 580–587. Springer Verlag, 1992.

[81] P. Kall and J. Mayer. SLP-IOR: On the design of a workbench for testing SLP codes. *Revista Investigacion Operacional*, 14:148–161, 1993.

[82] P. Kall and J. Mayer. SLP-IOR: A model management system for stochastic linear programming. In G. Hellwig, P. Kall, and P. Abel, editors, *Statistical Methods for Decision Processes*, pages 54–63. Daimler Benz AG, Stuttgart-Möhringen, 1994.

[83] P. Kall, A. Ruszczyński, and K. Frauendorfer. Approximation techniques in stochastic programming. In Y. Ermoliev and R.J-B. Wets, editors, *Numerical Techniques for Stochastic Optimization*, pages 33–64. Springer Verlag, 1988.

[84] P. Kall and D. Stoyan. Solving stochastic programming problems with recourse including error bounds. *Mathematische Operationsforschung und Statistik, Ser. Optimization*, 13:431–447, 1982.

[85] P. Kall and S. W. Wallace. *Stochastic programming*. John Wiley & Sons, 1994.

[86] E. Keller. GENSLP: A program for generating input for stochastic linear programs with complete fixed recourse. Manuscript, IOR, University of Zurich, 1984.

[87] J. E. Kelley. The cutting plane method for solving convex programs. *SIAM J. Industrial and Appl. Mathematics*, 8:703–712, 1960.

[88] W. J. Kennedy Jr. and J. E. Gentle. *Statistical computing*. Marcel Dekker, 1980.

[89] A. J. King. Stochastic programming problems: Examples from the literature. In Y. Ermoliev and R.J-B. Wets, editors, *Numerical Techniques for Stochastic Optimization*, pages 543–567. Springer Verlag, 1988.

[90] W. K. Klein Haneveld, L. Stougie, and M. H. van der Vlerk. On the convex hull of the simple integer recourse objective function. Research memorandum 516, IER, University of Groningen, 1993.

[91] H. Kleinmichel and H. Sadowski. Der verallgemeinerte RG-Algorithmus bei linearen Restriktionen, die Behandlung des Entartungsfalls und die Konvergenz des Verfahrens. *Beiträge zur Numerischen Mathematik*, 3:37–55, 1975. Leipzig.

[92] D. E. Knuth. *The art of computer programming Volume 2, Seminumerical algorithms*. Addison-Wesley Publ. Co., 1969.

[93] É. Komáromi. A dual method for probabilistic constrained problems. *Mathematical Programming Study*, 28:94–112, 1986.

[94] S. M. Kwerel. Most stringent bounds on aggregated probabilities of partially specified dependent probability systems. *Journal of the American Statistical Association*, 70:472–479, 1975.

[95] F. A. Lootsma. Ranking of non-linear optimization codes according to efficiency and robustness. In L. Collatz, G. Meinardus, and W. Wetterling, editors, *Konstruktive Methoden der finiten nichtlinearen Optimierung*, pages 157–178. Birkhäuser, 1980.

[96] F. V. Louveaux and Y. Smeers. Optimal ivestments for electricity generation: A stochastic model and a test-problem. In Y. Ermoliev and R.J-B. Wets, editors, *Numerical Techniques for Stochastic Optimization*, pages 445–453. Springer Verlag, 1988.

[97] F. V. Louveaux and M. van der Vlerk. Stochastic programming with simple integer recourse. *Mathematical Programming*, 61:301–325, 1993.

[98] D. G. Luenberger. *Linear and nonlinear programming*. Addison-Wesley Publ. Co., 1984.

[99] I. J. Lustig, J. M. Mulvey, and T. J. Carpenter. Formulating two-stage stochastic programs for interior point methods. *Operations Research*, 39:757–770, 1991.

[100] A. Madansky. Methods of solution of linear programs under uncertainty. *Operations Research*, 10:463–470, 1962.

[101] R. E. Marsten. The design of the XMP linear programming library. *ACM Transactions on Mathematical Software*, 7:481–497, 1981.

[102] K. Marti. Konvexitätsaussagen zum linearen stochastischen Optimierungsproblem. *Zeitschrift für Wahrscheinlichkeitstheorie und verw. Geb.*, 18:159–166, 1971.

[103] K. Marti. *Descent directions and efficient solutions in discretely distributed stochastic programs*. Springer Verlag, 1988.

[104] K. Marti and E. Fuchs. Computation of descent directions and efficient points in stochastic optimization problems without using derivatives. *Mathematical Programming Study*, 28:132–156, 1986.

[105] K. Marti and E. Fuchs. Rates of convergence of semi-stochastic approximation procedures for solving stochastic optimization problems. *Optimization*, 17:243–265, 1986.

[106] J. Mayer. Computational experiences with the reduced gradient method. In A. Prékopa, editor, *Progress in Operations Research*, pages 613–624. North-Holland Publ. Co., 1974.

[107] J. Mayer. A nonlinear programming method for the solution of a stochastic programming model of A. Prekopa. In A. Prékopa, editor, *Survey of Mathematical Programming, Vol. 2.*, pages 129–139. North-Holland Publ. Co., 1979.

[108] J. Mayer. Probabilistic constrained programming: A reduced gradient algorithm implemented on PC. Working Paper WP-88-39, IIASA, 1988.

[109] J. Mayer. Computational techniques for probabilistic constrained optimization problems. In K. Marti, editor, *Stochastic Optimization: Numerical Methods and Technical Applications*, pages 141–164. Springer Verlag, 1992.

[110] G. P. McCormick. *Nonlinear programming. Theory, Algorithms, and Applications.* John Wiley & Sons, 1983.

[111] L. McLinden. An analogue of moreaus proximation theorem, with application to the nonlinear complementarity problem. *Pacific Journal of Mathematics*, 88:101–161, 1980.

[112] Cs. Mészáros. The augmented system variant of IPMs in two–stage stochastic linear programming computation. Working Paper WP-95-11, MTA SzTAKI, Budapest, 1995.

[113] R. Meyer. The validity of a family of optimization methods. *SIAM Journal on Control*, 8:41–54, 1970.

[114] R.D. Monteiro and F. Zhou. On the existence and convergence of the central path for convex programming and some duality results. Technical report, Georgia Inst. of Technology, School of Op. Res., 1994.

[115] J. M. Mulvey, editor. *Evaluating mathematical programming techniques.* Springer Verlag, 1982.

[116] B. A. Murtagh. *Advanced linear programming: Computation and practice.* McGraw-Hill, 1981.

[117] B. A. Murtagh and M. A. Saunders. Large scale linearly constrained optimization. *Mathematical Programming*, 14:41–72, 1978.

[118] L. Nazareth and R. J-B. Wets. Nonlinear programming techniques applied to stochastic programs with recourse. In Y. Ermoliev and R.J-B. Wets, editors, *Numerical Techniques for Stochastic Optimization*, pages 95–121. Springer Verlag, 1988.

[119] G. L. Nemhauser and W. B. Widhelm. A modified linear program for columnar methods in mathematical programming. *Operations Research*, 19:1051–1060, 1971.

[120] J. Pfanzagl. Convexity and conditional expectations. *Annals of Probability*, 2:490–494, 1974.

[121] A. Prékopa. Logarithmic concave measures with applications to stochastic programming. *Acta. Sci. Math*, 32:301–316, 1971.

[122] A. Prékopa. A class of stochastic programming decision problems. *Mathematische Operationsforschung und Statistik*, 3:349–354, 1972.

[123] A. Prékopa. Contributions to stochastic programming. *Mathematical Programming*, 4:202–221, 1973.

[124] A. Prékopa. Eine Erweiterung der sogenannten Methode der zulässigen Richtungen der nichtlinearen Optimierung auf den Fall quasikonkaver Restriktionen. *Mathematische Operationsforschung und Statistik, Ser. Optimization*, 5:281–293, 1974.

[125] A. Prékopa. Logarithmic concave measures and related topics. In M. A. H. Dempster, editor, *Stochastic Programming*, pages 63–82. Academic Press, 1980.

[126] A. Prékopa. Numerical solution of probabilistic constrained programming problems. In Y. Ermoliev and R.J-B. Wets, editors, *Numerical Techniques for Stochastic Optimization*, pages 123–139. Springer Verlag, 1988.

[127] A. Prékopa. Boole-bonferroni inequalities and linear programming. *Operations Research*, 36:145–162, 1988.

[128] A. Prékopa. Sharp bounds on probabilities using linear programming. *Operations Research*, 38:227–239, 1990.

[129] A. Prékopa. A dual method for the solution of a one-stage stochastic programming problem with random RHS obeying a discrete probability distribution. *Zeitschrift für Operations Research*, 34:441–461, 1990.

[130] A. Prékopa. *Stochastic programming.* Kluwer Academic Publishers, 1995.

[131] A. Prékopa, S. Ganczer, I. Deák, and K. Patyi. The STABIL stochastic programming model and its experimental application to the electricity production in hungary. In M. A. H. Dempster, editor, *Stochastic Programming*, pages 369–385. Academic Press, 1980.

[132] A. Prékopa and P. Kelle. Reliability type inventory models based on stochastic programming. *Mathematical Programming Study*, 9:43–58, 1983.

[133] T. Rapcsák. *On the numerical solution of a reservoir model.* PhD thesis, University of Debrecen, 1974. in Hungarian.

[134] Y. Rinott. On convexity of measures. *Annals of Probability*, 4:1020–1026, 1976.

[135] A. Rényi. *Probability theory.* North-Holland Publ. Co., 1970.

[136] R.T. Rockafellar. *Convex analysis.* Princeton University Press, 1970.

[137] R.T. Rockafellar. Monotone operators and the proximal point algorithm. *SIAM Journal of Control and Optimization*, 14:877–898, 1976.

[138] A. Ruszczyński. A regularized decomposition method for minimizing a sum of polyhedral functions. *Mathematical Programming*, 35:309–333, 1986.

[139] A. Ruszczyński. A linearization method for nonsmooth stochastic programming problems. *Mathematics of Operations Research*, 12:32–49, 1987.

[140] A. Ruszczyński. Interior point methods in stochastic programming. Working Paper WP-93-8, IIASA, 1993.

[141] A. Ruszczyński. Regularized decomposition of stochastic programs: algorithmic techniques and numerical results. Working Paper WP-93-21, IIASA, 1993.

[142] A. Ruszczyński and A. Świętanowski. On the regularized decomposition method for two stage stochastic linear problems. Working Paper WP-96-014, IIASA, 1996.

[143] H. Sadowski. *Untersuchungen zu den Verfahren der reduzierten Gradienten in der nichtlinearen Optimierung.* PhD thesis, University of Technology, Dresden, 1973.

[144] K. Schittkowski. *Nonlinear programming codes - Information, tests and performance.* Springer Verlag, 1980.

[145] K. Schittkowski. EMP: An expert system for mathematical programming. Technical report, Mathematisches Institut, Universität Bayreuth, 1987.

[146] K. Schittkowski. Some experiments on heuristic code selection versus numerical performance in nonlinear programming. *European Journal of Operational Research*, 65:292–304, 1993.

[147] Gy. Sonnevend. An analytical centre for polyhedrons and new classes for linear (smooth, convex) programming. In A. Prékopa et al., editors, *System Modelling and Optimization*, pages 866–878. Springer Verlag, 1986.

[148] M. Stefik and D. G. Bobrow. Object-oriented programming: Themes and variations. *The AI Magazine*, 6:40–62, 1986.

[149] B. Strazicky. On an algorithm for solution of the two-stage stochastic programming problem. *Methods of Operations Research*, XIX:142–156, 1974.

[150] T. Szántai. Evaluation of a special multivariate gamma distribution. *Mathematical Programming Study*, 27:1–16, 1986.

[151] T. Szántai. A computer code for solution of probabilistic-constrained stochastic programming problems. In Y. Ermoliev and R.J-B. Wets, editors, *Numerical Techniques for Stochastic Optimization*, pages 229–235. Springer Verlag, 1988.

[152] T. Szántai. Calculation of the multivariate probability distribution function values and their gradient vectors. Working Paper WP-87-82, IIASA, 1987.

[153] L. Takács. On the method of inclusion and exclusion. *Journal of the American Statistical Association*, 62:102–113, 1967.

[154] J. Tind and L.A. Wolsey. An elementary survey of general duality theory in mathematical programming. *Mathematical Programming*, 21:241–261, 1981.

[155] A. F. Veinott. The supporting hyperplane method for unimodal programming. *Operations Research*, 15:147–152, 1967.

[156] R. Van Slyke and R. J-B. Wets. L–shaped linear program with applications to optimal control and stochastic linear programs. *SIAM J. Appl. Math.*, 17:638–663, 1969.

[157] M. H. van der Vlerk. *Stochastic programming with integer recourse*. Labyrint Publ. Capelle aan den IJssel, The Netherlands, 1995.

[158] R. J-B. Wets. Solving stochastic programs with simple recourse. *Stochastics*, 10:219–242, 1983.

[159] R. J-B. Wets. Stochastic programming: Solution techniques and approximation schemes. In A. Bachem, M. Grötschel, and B. Korte, editors, *Mathematical programming: The state of the art*, pages 566–603. Springer Verlag, 1983.

[160] R. J-B. Wets. Large scale linear programming techniques. In Y. Ermoliev and R.J-B. Wets, editors, *Numerical Techniques for Stochastic Optimization*, pages 65–93. Springer Verlag, 1988.

[161] R. J-B. Wets. Stochastic programming. In G. L. et al. Nemhauser, editor, *Handbooks in OR and MS, Vol. 1*, pages 573–629. Elsevier, 1989.

[162] P. Wolfe. Methods of nonlinear programming. In R. L. Graves and P. Wolfe, editors, *Recent results in mathematical programming*, pages 67–86. McGraw-Hill Publ. Co., 1963.

[163] P. Wolfe. Convergence of gradient methods under constraint. *IBM Journal of Research and Development*, 16:407–411, 1972.

[164] L.A. Wolsey. A resource decomposition algorithm for general mathematical programs. *Mathematical Programming Study*, 14:244–257, 1981.

[165] G. Zoutendijk. *Methods of feasible directions*. Elsevier Publ. Co., 1960.

[166] G. Zoutendijk. Nonlinear programming: A numerical survey. *J. SIAM Control*, 4:194–210, 1966.

Author Index

Subject Index

barrier methods, 40–42, 80
   logarithmic barrier, 41, 80
      analytic center, 42
      central path, 42
Benders decomposition, 3, 16, **17**, 69, 74, 77, 88
   dual block–angular problem
      aggregate cuts, *see* L–shaped method
      disaggregate cuts, *see* multicut method
binomial moments, 66, 67
Boole–Bonferroni inequalities, 66, 67, 81
BPMPD, 69

candidate solution, *see* regularized outer approximation
central cutting plane methods
   Elzinga–Moore method, 43
   general scheme, 43
   Goffin–Vial method, 43
   modified Elzinga–Moore method, **44**, 82
chance constrained model, *see* stochastic programming model
chance constrained models, algorithms, 80–84
   central cutting plane method, 82, 89
      implementation, 89
   finding a Slater point, 84
   reduced gradient method, 82, 89
      implementation, 89
   supporting hyperplane method, 81, 89
      implementation, 89
chance constraints
   joint, 59

For index entries with several page numbers boldface indicates definition or specification of the corresponding notion.

separate, 59
class, *see* object–oriented style
computational environment, 92
convex programming
   duality, 1–6
      duality theorem, 2
      Lagrange dual, **2**, 3, 15
      Lagrange multipliers, **2**, 34, 35, 40
      Tind–Wolsey dual, **3**, 6
   Kuhn–Tucker conditions, 4, 103
   Kuhn–Tucker theorem, **4**, 46
   parametric, 4
   regularity conditions, 2
      Slater regularity, **2**, 4, 12–15, 22, 31, 43–45, 52, 80–82, 84, 89, 103, 128
correlation matrix, 102
cut
   feasibility, **10**, 12, 15, 20, 21, 26, 28, 30–32, 34, 37–39, 43, 44
   incumbent, 78
   Kelley's, **12**, 43, 44
   objective (central), 43, 45
   optimality, **10**, 20, 21, 28, 31, 35, 37–39
   Veinott's, **13**, 44
cutting plane methods
   Kelley's method, 12
   Veinott's method, **13**, 81

DAPPROX, **86**, 87, 106, 113–126, 134
DECOMP, 86
direction
   dual ray, **3**, 6, 14, 15, 17
   extremal, 16, 17
   feasible, 28, 46, 48
discrete approximation methods, *see* suc-